From Small Organic Molecules to Large

A Century of Progress

From Small Organic Molecules to Large

A Century of Progress

Herman F. Mark

PROFILES, PATHWAYS, AND DREAMS
Autobiographies of Eminent Chemists

Jeffrey I. Seeman, Series Editor

American Chemical Society, Washington, DC 1993

Library of Congress Cataloging-in-Publication Data

Mark, H. F. (Herman Francis)1895–1991
From small organic molecules to large: a century of progress/ Herman F. Mark
p. cm.—(Profiles, pathways, and dreams, ISSN 1047–8329)
Includes bibliographical references and index.
ISBN 0–8412–1776–9.—ISBN 0–8412–1802–1 (pbk.)
1. Mark, H. F. (Herman Francis), 1895–1991.
2. Chemists—United States—Biography. 3. Chemistry, Organic—United States—History—20th century.
4. Polymers. I. Title. II. Series.

QA29.M358A3 1993
540'.92 92–30099
[B] CIP

Jeffrey I. Seeman, Series Editor

The paper used in this publication meets the minimum requirements of American National Standard for Information Sciences—Permanence of Paper for Printed Library Materials, ANSI Z39.48–1984.

(∞)

PRINTED IN THE UNITED STATES OF AMERICA

1993 Advisory Board

Foreword

In 1986, the ACS Books Department accepted for publication a collection of autobiographies of organic chemists, to be published in a single volume. However, the authors were much more prolific than the project's editor, Jeffrey I. Seeman, had anticipated, and under his guidance and encouragement, the project took on a life of its own. The original volume evolved into 22 volumes, and the first volume of Profiles, Pathways, and Dreams: Autobiographies of Eminent Chemists was published in 1990. Unlike the original volume, the series was structured to include chemical scientists in all specialties, not just organic chemistry. Our hope is that those who know the authors will be confirmed in their admiration for them, and that those who do not know them will find these eminent scientists a source of inspiration and encouragement, not only in any scientific endeavors, but also in life.

M. Joan Comstock
Head, Books Department
American Chemical Society

Contributors

We thank the following corporations and Herchel Smith for their generous financial support of the series Profiles, Pathways, and Dreams.

Akzo nv

Bachem Inc.

E. I. du Pont de Nemours
and Company

Duphar B.V.

Eisai Co., Ltd.

Fujisawa Pharmaceutical Co., Ltd.

Hoechst Celanese Corporation

Imperial Chemical Industries PLC

Kao Corporation

Mitsui Petrochemical Industries,
Ltd.

The NutraSweet Company

Organon International B.V.

Pergamon Press PLC

Pfizer Inc.

Philip Morris

Quest International

Sandoz Pharmaceuticals
Corporation

Sankyo Company, Ltd.

Schering–Plough Corporation

Shionogi Research Laboratories,
Shionogi & Co., Ltd.

Herchel Smith

Suntory Institute for Bioorganic
Research

Takasago International
Corporation

Takeda Chemical Industries, Ltd.

Unilever Research U.S., Inc.

Profiles, Pathways, and Dreams

Titles in This Series

Sir Derek H. R. Barton *Some Recollections of Gap Jumping*

Arthur J. Birch *To See the Obvious*

Melvin Calvin *Following the Trail of Light: A Scientific Odyssey*

Donald J. Cram *From Design to Discovery*

Michael J. S. Dewar *A Semiempirical Life*

Carl Djerassi *Steroids Made It Possible*

Ernest L. Eliel *From Cologne to Chapel Hill*

Egbert Havinga *Enjoying Organic Chemistry, 1927–1987*

Rolf Huisgen *The Adventure Playground of Mechanisms and Novel Reactions*

William S. Johnson *A Fifty-Year Love Affair with Organic Chemistry*

Raymond U. Lemieux *Explorations with Sugars: How Sweet It Was*

Herman Mark *From Small Organic Molecules to Large: A Century of Progress*

Bruce Merrifield *Life During A Golden Age of Peptide Chemistry*

Teruaki Mukaiyama *To Catch the Interesting While Running*

Koji Nakanishi *A Wandering Natural Products Chemist*

Tetsuo Nozoe *Seventy Years in Organic Chemistry*

Vladimir Prelog *My 132 Semesters of Chemistry Studies*

John D. Roberts *The Right Place at the Right Time*

Paul von Rague Schleyer *From the Ivy League into the Honey Pot*

F. G. A. Stone *Leaving No Stone Unturned*

Andrew Streitwieser, Jr. *A Lifetime of Synergy with Theory and Experiment*

Cheves Walling *Fifty Years of Free Radicals*

About the Editor

JEFFREY I. SEEMAN received his B.S. with high honors in 1967 from the Stevens Institute of Technology in Hoboken, New Jersey, and his Ph.D. in organic chemistry in 1971 from the University of California, Berkeley. Following a two-year staff fellowship at the Laboratory of Chemical Physics of the National Institutes of Health in Bethesda, Maryland, he joined the Philip Morris Research Center in Richmond, Virginia. In 1983–1984, he enjoyed a sabbatical year at the Dyson Perrins Laboratory in Oxford, England, and claims to have visited more than 90% of the castles in England, Wales, and Scotland.

Seeman's 90 published papers include research and patents in the areas of photochemistry, nicotine and tobacco alkaloid chemistry and synthesis, conformational analysis, pyrolysis chemistry, organotransition metal chemistry, the use of cyclodextrins for chiral recognition, and structure–activity relationships in olfaction. He was a plenary lecturer at the Eighth IUPAC Conference on Physical Organic Chemistry held in Tokyo in 1986 and has been an invited lecturer at numerous scientific meetings and universities. Currently, Seeman serves on the Petroleum Research Fund Advisory Board. He continues to count Nero Wolfe and Archie Goodwin among his best friends.

Contents

Photographs

Preface

"HOW DID YOU GET THE IDEA—and the good fortune—to convince 22 world-famous chemists to write their autobiographies?" This question has been asked of me, in these or similar words, frequently over the past several years. I hope to explain in this preface how the project came about, how the contributors were chosen, what the editorial ground rules were, what was the editorial context in which these scientists wrote their stories, and the answers to related issues. Furthermore, several authors specifically requested that the project's boundary conditions be known.

As I was preparing an article[1] for *Chemical Reviews* on the Curtin–Hammett principle, I became interested in the people who did the work and the human side of the scientific developments. I am a chemist, and I also have a deep appreciation of history, especially in the sense of individual accomplishments. Readers' responses to the historical section of that review encouraged me to take an active interest in the history of chemistry. The concept for Profiles, Pathways, and Dreams resulted from that interest.

My goal for Profiles was to document the development of modern organic chemistry by having individual chemists discuss their roles in this development. Authors were not chosen to represent my choice of the world's "best" organic chemists, as one might choose the "baseball all-star team of the century". Such an attempt would be foolish: Even the selection committees for the Nobel prizes do not make their decisions on such a premise.

The selection criteria were numerous. Each individual had to have made seminal contributions to organic chemistry over a multidecade career. (The average age of the authors is over 70!) Profiles would represent scientists born and professionally productive in different countries. (Chemistry in 13 countries is detailed.) Taken together, these individuals were to have conducted research in nearly all subspecialties of organic chemistry. Invitations to contribute were based on solicited advice and on recommendations of chemists from five continents, including nearly all of the contributors. The final assemblage was selected entirely and exclusively by me. Not all who were invited chose to participate, and not all who should have been invited could be asked.

A very detailed four-page document was sent to the contributors, in which they were informed that the objectives of the series were

1. to delineate the overall scientific development of organic chemistry during the past 30–40 years, a period during which this field has dramatically changed and matured;
2. to describe the development of specific areas of organic chemistry; to highlight the crucial discoveries and to examine the impact they have had on the continuing development in the field;
3. to focus attention on the research of some of the seminal contributors to organic chemistry; to indicate how their research programs progressed over a 20–40-year period; and
4. to provide a documented source for individuals interested in the hows and whys of the development of modern organic chemistry.

One noted scientist explained his refusal to contribute a volume by saying, in part, that "it is extraordinarily difficult to write in good taste about oneself. Only if one can manage a humorous and light touch does it come off well. Naturally, I would like to place my work in what I consider its true scientific perspective, but . . ."

Each autobiography reflects the author's science, his lifestyle, and the style of his research. Naturally, the volumes are not uniform, although each author attempted to follow the guidelines. "To write in good taste" was not an objective of the series. On the contrary, the authors were specifically requested not to write a review article of their field, but to detail their own research accomplishments. To the extent that this instruction was followed and the result is not "in good taste", then these are criticisms that I, as editor, must bear, not the writer.

As in any project, I have a few regrets. It is truly sad that Herman Mark, who wrote this volume, Egbert Havinga, who wrote another, and David Ginsburg, who translated Vladimir Prelog's volume, died during the course of this project. There have been many rewards, some of which are documented in my personal account of this project, entitled "Extracting the Essence: Adventures of an Editor" published in *CHEMTECH*.[2]

Acknowledgments

I join the entire scientific community in offering each author unbounded thanks. I thank their families and their secretaries for their contributions. Furthermore, I thank numerous chemists for reading and reviewing the autobiographies, for lending photographs, for sharing information, and for providing each of the authors and me the encouragement to proceed in a project that was far more costly in time and energy than any of us had anticipated.

I thank my employer, Philip Morris USA, and J. Charles, R. N. Ferguson, H. Grubbs, K. Houghton, and W. F. Kuhn, for without their support Profiles, Pathways, and Dreams could not have been. I thank ACS Books, and in particular, Robin Giroux (production manager), Janet Dodd (senior editor), Joan Comstock (department head), and their staff for their hard work, dedication, and support. Each reader no doubt joins me in thanking 24 corporations and Herchel Smith for financial support for the project.

I thank my children, Jonathan and Brooke, for their patience and understanding; remarkably, I have been working on Profiles for more than half of their lives—probably the only half that they can remember! Finally, I again thank all those mentioned and especially my family, friends, colleagues, and the 22 authors for allowing me to share this experience with them.

JEFFREY I. SEEMAN
Philip Morris Research Center
Richmond, VA 23234

March 26, 1993

[1] Seeman, J. I. *Chem. Rev.* **1983,** *83*, 83–134.
[2] Seeman, J. I. *CHEMTECH* **1990**, *20*(2), 86–90.

Editor's Note

THE PHONE ON MY DESK RANG. "Hello!HELLO!HELLO!" his warm, strong voice greeted me. "Howareyou?HOWAREYOU?HOWAREYOU?" Just hearing Herman Mark's salutation sparked good cheer. I can hear the reverberations still.

Our last telephone conversation, in early February of 1992, began in just such fashion. He was calling from Texas where, due to failing health, he was residing with his elder son, Hans. "I hope to return to Brooklyn in the spring, when it gets warmer," Mark said. "Thank you very much! THANKYOU!THANK YOU! And most important, Happy Valentine's Day!" It was typical Mark gusto, sadly a bit weaker than I remembered from previous calls. Little did I know that it was to be our last conversation. Mark passed away on April 6 of that year, just shy of his 97th birthday.

Among the last batch of comments Mark had written for his book was a possible alternative title, "My Life: One Hundred Years of Random Walk". "I do not travel," Mark had commented to me a few years previously. "I am *being* traveled. It is a form of Brownian motion!" Mark's travels took him to four continents and to uncharted areas in science and technology. Born in Vienna, he became the most highly decorated officer in the Austrian army in World War I. After the war, he returned to Vienna to obtain his doctorate, then continued his travels: to Berlin as an instructor, then Dahlem (the Kaiser Wilhelm Institute) and Ludwigshafen (I. G. Farben), all in Germany. As Nazi oppression and anti-Semitism in Germany increased, Mark's job and life were threatened. He returned home to the University of Vienna to serve as Director of the Chemical Institute, hoping that he and his family would be safe. They weren't, and immediately following a brief but horrifying imprisonment at the hands of the Gestapo, Mark made his true escape, first to Canada and then to the United States. Brooklyn became his permanent home, although he traveled extensively, including frequent long visits to Vienna.

Mark's long life gave him many opportunities, not merely to travel but to interact with people around the world. At his 95th birthday celebration Mark had commented, "I have had the good fortune to participate in many areas of science for almost an entire century. It has been an amazing experience; our century has seen astounding intellectual progress. . . ."

As his son Hans so aptly put it, "My father's greatest accomplishment was as a pioneer in applying modern physics to chemistry." Mark's first fundamental contributions to science were multidisciplinary. In 1930, he and Raimund Wierl initiated the field of fast electron diffraction of gaseous molecules. This technique and knowledge, later pursued and advanced by Linus Pauling and others, led to thousands of important quantitative structural determinations.

Mark's second intellectual life dealt with the synthesis, characterization, properties, and reactivity of polymers—natural and synthetic. He founded the world's first comprehensive academic curriculum of polymer technology at the University of Vienna in 1932. But it was at Brooklyn's Polytechnic Institute that Mark truly earned his title of a "founding father" of polymer science and technology. Mark established the first institute in the United States dedicated to polymers. He became a prolific writer and editor, establishing the prestigious *Journal of Polymer Science* and *Journal of Applied Polymer Science* in addition to publishing some 600 articles and 20 books! One of his colleagues has called him "a mover and a shaker of science."

It is important to recall that polymer science was not yet prestigious when Mark joined the Brooklyn Poly faculty in 1940. Herbert Morawetz recalls:

> When I joined the Polymer Research Institute in 1951, Herman Mark suggested that I take over his graduate student Riad Gobran. Riad had a B.S. degree from the University of Alexandria, Egypt, and I was puzzled why he had come to Poly. When I asked him, he brought me a letter he had received from Professor C. Gardner Swain of M.I.T. in response to his application. It read, "We at M.I.T. specialize in the classical disciplines of organic, inorganic, and physical chemistry. Since you intend to restrict yourself to the

chemistry of polymers, I suggest that you apply to the Polytechnic Institute of Brooklyn." This was typical of the attitude of the chemistry establishment at that time.

According to Mark's former student Edward Teller, "Mark was probably the best lecturer I have ever met. His lectures could have been printed just as they were given." Mark also held Teller in high regard, commenting that Teller "often clarified the leading ideas for the class and myself." Interestingly, Mark's son Hans was a Teller student, so according to Mark and Teller, the latter "became the link in the generation gap of the Mark family."

Mark combined his ability as an excellent lecturer with his broad knowledge base, increased by his visits to various universities and industries. He made early connections between academic and industrial research, years before such ties became widely recognized as essential to human progress. For nearly 25 years, Mark presented at Poly the result of his travels in a lecture entitled "What's New in Polymers". According to one of his Poly colleagues, "we all looked forward to this masterpiece. Mark had a remarkable feeling as to what would become important. He was a catalyst for us." Further proof of Mark's greatness was the many famous scientists who were his students: to name a few, in addition to Teller, were Eugene Wigner, Odd Hassell, Leo Szilard, and Max Perutz.

Mark had an eye for fun and loved celebrations, especially birthday parties, which become more and more special with the years. For his 85th birthday, there were seven major celebrations (one each in New York, Philadelphia, Chicago, Moscow, and Germany; and two in Israel). "My birthday is like the Saltzburg Music Festival," he quipped. "It starts in April and ends in October."

"I am a European co-polymer," Mark once joked with me. I interpreted that to mean that Mark pulled together, for his own model, the best of all the people he met. Mark's dominant personality trait was optimism. He had the amazing capability to see the good side of everything and of anyone. He would not allow anything to come in the way of positive human interactions. His door was always open.

Mark certainly did not hold grudges. In reference to the Nazis and Mark's escape from his beloved Vienna, his son Hans summarized Mark's beliefs: "My father did not resent the past. This was his conscious decision; we discussed it as a family. He was determined to move on. He had an unquenchable optimism." Mark's decision to have his ashes buried beside his wife's grave in Vienna attests to his love of his home country and his unwillingness to judge the future by the past.

In the 1930s, Mark suffered many human, political, and scientific indignities. The interactions with Hermann Staudinger are famous.

Staudinger believed that polymers were rigid, "rod-like" species; Mark felt that polymers had significant conformational flexibility. Staudinger took this as a personal affront. Mark once sent Staudinger a gift of one of his books. It was returned, marked "unopened". And Herbert Morawetz recalls that when visiting Professor Staudinger's widow, he saw another of Mark's books flagged with a note in the Professor's handwriting stating, "Not science, propaganda." When Staudinger visited the United States in the early 1950s, he was amazed to received Mark's good will and hospitality.

Mark liked to communicate and to teach, but he did not like to grade students. Early in his career at Poly, he gave A's to everyone! This was not to the dean's liking; Mark's solution: grade everyone on a repeating cycle of two A's and one B grade. Morawetz recalls his own disappointment in receiving a grade of a B. "I went to see the mighty professor, asking what I had done wrong. Mark responded jovially that he would be glad to give me an oral examination. I got cold feet and left. Only much later did I learn Mark's grading scheme!"

One anecdote summarizes Mark's personal energy and memory. Again, as told by Morawetz:

> After I completed my studies, Mark asked whether I should like to join the Polymer Research Institute. He suggested that I visit him to talk things over. When I arrived, he was in bed with the flu. Had I not lived in Prague? he asked me. If so, I would certainly enjoy hearing the recording of a song "Die Novaks aus Prag", which he had just received from Vienna. I saw the *Geheimrat* for the last time a year before his death when I visited him in Texas. He was in a wheelchair, but planned to lecture on conducting polymers and had a lot of literature to prepare his talk [see seminar notice on page xxvi]. When I reminded him of my visit in 1951, he rolled his wheelchair over to a stack of gramophone records and played for me again "Die Novaks aus Prag".

Special Note

I am very sad that Mark did not live to see this completed volume. He worked so hard on it. Indeed, during the early writing stages, I could not keep pace with the man. I thought, "What a feeling to be running after a 93-year-old scientist!"

We owe a debt of gratitude to Herbert Morawetz who not only contributed wonderful anecdotes about his old friend, but who carefully and thoughtfully provided invaluable editing and proofreading support as the book progressed toward publication.

JEFFREY I. SEEMAN
Philip Morris Research Center
Richmond, VA 23234

March 26, 1993

Center for Polymer Research
The University of Texas at Austin

is pleased to announce the first in a Series
of Informal Lectures and Discussions

by

Professor Herman Mark

on

POLYMERIC MATERIALS OF THE FUTURE:
Lessons from Nature and History

Friday at 2:00 p.m.
August 24, 1990.

Chemical and Petroleum Engineering Bldg.,
Room 2.222

Even at age 95, Mark was active and vigorous! It is interesting that he would give a series of lectures on science of the future and not on lessons learned from the past. I wish I had had the opportunity to attend the series.

From Small Organic Molecules to Large

A Century of Progress

Herman Mark

Childhood, Adolescence, and World War I

1895–1918

My father, Herman Carl Mark, was born in Hungary in 1861. At the time of my birth (1895) he was an assistant surgeon at a hospital in the fourth district of Vienna. He was a well-liked practitioner because he came from the famous Viennese School of Medicine and spoke four languages (German, Hungarian, Czech, and Polish) fluently. He talked in simple terms, a trait that helped him greatly with his patients. Our family—father, mother Lili (neé Mueller), my brother Hans (who was 1 year younger than I), my sister Elizabeth (10 years younger), and I—were reasonably well-to-do.

I entered public school in 1900, started high school in 1905, and graduated in 1913. Remembering these 13 years of childhood and adolescence, I can now see that they were a most unusual and attractive mixture of discipline and freedom. My parents insisted that we take proper care in our school work, a task that was relatively easy at the beginning but increasingly difficult and absorbing in the higher grades. Both parents were very strict about poor marks and late arrival at school, and any kind of negligence was not accepted but usually punished very rapidly. There was no hard beating, but many slaps were dished out and absorbed. In the rarefied atmosphere of hindsight, I

In July 1913, I graduated from the Teresianische Akademic high school as one of a class of 19 (second from left, back row). A few others later became prominent scientists and politicians, i.e., Ressig, Riehl, Coudenhove (the indefatigable propagandist for a united Europe), and Toldt. During our lives we kept our friendship and respect for each other.

confess that it did not do us any harm, only good. As Hans recalled* those days:

> My brother always had high grades except in behavior, and this was a great pain to our father. One day he came home early and told our mother that he had heard one gentleman tell another: "The worst boys in the area are the Mark boys!"[1]

Once the work was done acceptably, there was complete freedom, and the "sky was the limit". And what a sky it was—studded with many alluring types of sparkling stars.

**The Series Editor has selected material that he felt would expound upon the information provided by Dr. Mark from ACS Symposium Series 175;* Polymer Science Overview: A Tribute to Herman F. Mark; *Stahl, G. Allan, Ed. American Chemical Society: Washington, DC; 1981. Material from this book will be found throughout the volume.*

We participated in sports and physical exercises of all kinds: skating, soccer, and long extended skiing tours during the winter; tennis and swimming in the Danube in the summer; and sailing and mountain climbing during vacations. In all these disciplines, my brother, a few friends, and I reached a level of skill that was good enough to provide fun and pleasure without attempting to reach a championship. There would not have been enough time or money for that. Still, on one occasion I participated in a game on the Austrian National Soccer Team.

At the age of 13, on vacation in a small Tyrolean village. The life of my brother and myself was very pleasant and free, with sports such as hiking in the hills, climbing difficult peaks, and swimming in a nearby lake. Very important, also, were other sports: soccer and tennis in Vienna, skiiing and cycling in the country.

A gifted athlete, Mark ran to and from school, often exhausting his friends with seeming unending energy. He played on a local football club, and became an accomplished skier and mountain climber. Besides the obvious physical maturation sports nourished, there was an additional benefit. Through sports, an attitude of healthy competition was developed. Not as established as today, sports of that era stressed physical exercise and competition of skills. Winning was not as important as it is now.[1]

Music

Another inexhaustible fountain of relaxation and pleasure was music, both at home and in public. My father, coming from Hungary, was a competent fiddler, and my mother, hailing from Vienna, had acquired some skill on the piano as a child. Both were musical and music-loving. Once or twice a week they played one of the classical, not too complicated, violin–piano sonatas of Haydn, Mozart, Schubert, and eventually even

As a high school soccer player in Vienna, 1912.

Beethoven. Whether we children were still up or already in bed, we listened breathlessly and inhaled the serenity and beauty of these immortal tunes that were to enrich the rest of our lives.

The performances in the music halls and operas made a still deeper and longer-lasting impression. Vienna was at that time (1880–1900) the world metropolis of music. There lived and worked in Vienna Johannes Brahms, Anton Bruckner, Gustav Mahler, Richard Strauss, Johann Strauss, and Wilhelm Krenzl. There were many famous conductors such as Franz Schalk, Bruno Walter, and Hans Richter. Two of the best orchestras—the philharmonic and symphonic—complemented a galaxy of preeminent soloists on piano, violin, cello, and most other solo instruments.

Listening to music at home was easy and did not require much time. On the other hand, a ticket for a first-class performance of the opera required standing in a line for several hours. We solved this problem by constructing very light chairs that we could take with us into the theater. Then, sitting for 3 hours on these makeshift chairs, we were able to complete our school work for the next day conveniently and perfectly before listening to the opera. We listened to all of them—classical, romantic, modern, and super-modern (at that time!)— over and over again until we knew many of them almost by heart. The visits to symphonies, solo concerts, recitals, and ballets were similar but required less effort because it was much easier to get tickets and the waiting lines were much shorter.

There was also time for tragic and comic plays by Shakespeare, Molière, Schiller, Ibsen, and many other classical and modern authors. Other arts, such as paintings, drawings, sculptures, and jewelry, were abundant in the 20-odd museums and even more numerous churches of the city.

Altogether, this theater-going led to a deep and lasting appreciation of the cultural value of the European tradition since the Middle Ages. It highlighted the fundamentals of human creative progress: imagination, sustained effort, skill, precision, and endless patience. All of this made a deep impression on youngsters in their formative years; I wish we would have more of it today. And it was available to the middle-class society that made up a large percentage of the population. Our family and our school were not exceptions; this outlook was the *rule*.

Academic Atmosphere

Another important influence on my education and on that of my classmates was our teacher in mathematics and physics, Franz Hlawaty. He was an educational genius with these disciplines that are usually hated for being "difficult" and "abstract". He patiently and carefully explained to us not only the principles but also their applicability. Suddenly everything made sense; the lessons were no longer irrational, but useful. I have no doubt in my mind that Dr. Hlawaty influenced me greatly to select science for my career.

Quite different factors between 1905 and 1913 also greatly influenced me and my later decisions. One of my close friends was Gerhardt Kirsch, 4 years older than I, who later became a distinguished professor of physics at the University of Vienna. He graduated from our high school in 1909 and entered the university as a physics student in the same year. At that time I was still a high school boy and could not enter the holy halls of the university, but Gerhardt "smuggled" me in with him, and in this way I got an early and very exciting glimpse of university life. Gerhardt and I met often throughout life, and I always profited from these meetings.

> Herman and Gerhardt's interest in science was furthered by one of their teachers, a priest named Hlawaty. Father Hlawaty taught mathematics, physics, and chemistry. With his lectures, and especially his demonstrations and experiments, he piqued their imaginations, and, according to Mark, was the one man who most interested them in the study of nature and, in particular, physical science.
>
> . . . Mark and his close friend, Gerhardt Kirsch, toured the laboratories of the University of Vienna at the invitation of Gerhardt's father, a professor of material science. The boys returned home excited by "the sight and flasks and beakers and retorts, of blue flamed Bunsen burners, of bubbling liquids and the lengths of rubber tubing that carried off their vapors". The cliché is probably accurate in so far as "nothing was ever the same again".[1]

The famous and unforgettable Ludwig Boltzmann, a professor of physics in Vienna and brilliant creator of statistical mechanics, had died only a few years earlier. It was he who provided the connection between classical thermodynamics and modern atomic theory. His memory was kept very vividly alive by some of his followers and students such as Stefan Meyer, Fritz Hasenöhrl, Karl Przibram, and Lise Meitner, all of whom later became famous and preeminent scientists, educators, and school principals.

During the last years before the war (1910–1914), many internationally famous scientists came to the University of Vienna to deliver lectures and to hold seminars. Gerhardt took me to some of them so that I met at an early age such scientific giants as Emil Fischer from Berlin, Albert Einstein from Prague, Ernest Rutherford from England, and Marie Curie from Paris—all of whom made deep and unforgetable impressions on a young mind.

Another interesting and, later, exciting "outside" influence came from my father's medical studies. For several years he was a classmate and colleague of Sigmund Freud. After they both had graduated, they remained good friends, and I remember that several alumni of the Old Viennese Medical School (Tandler, Schnitzler, Freud, and my father) had dinner in our house, where my mother, an excellent cook, took care of them. Most of my father's friends were Jewish, and a few of them were even Zionists (Otto Weich and Wilhelm Korwin). During those years (1912–1914), I heard for the first time the names Theodor Herzl and Chaim Weizmann; even the names conveyed the feeling that new and important ideas were gaining ground and significance not in science but in human affairs, culture and politics.

Two of my uncles, Otto Waldstein and Wilhelm Kraus, were members of the "Writers Club" and on very good terms with leading authors of those years such as Franz Werfel, Peter Altenberg, Hugo von Hofmannsthal, and Karl Kraus. I saw and heard all of them quarreling and arguing with each other in one of the famous Viennese literary coffee shops, Kaffee Zentral or Kaffee Griensteidl.

Military Service

In this manner, under the influence of many factors coming from all corners of human life, I grew up and reached my high school graduation in July 1913. At that point I had two options: to enlist in the army immediately as a volunteer for 1 year of service or to postpone this enlistment for 4 or 5 years in order to obtain a Ph.D. or M.D. I wanted to get past all military obligations and commitments, therefore, I enlisted immediately to enter service in the fall of 1913, intending to start my university studies in the fall of 1914.

Unfortunately, not only for me and my classmates but for all humanity, an unexpected terrible event took place. World War I broke out in July 1914 and brought with it 4 years of catastrophic events, misery, and suffering. I spent most of the war on the front line and was wounded three times, but with a good deal of luck, I survived. I earned 15 medals, including some of the highest awards. I received one of my decorations for "disobeying orders" to retreat. After the Italians had captured Monte Ortigora, I received permission to lead an attempt to recapture this strategic peak and we succeeded in doing so despite heavy losses.

At the end of the war, in November 1918, I was captured on the Italian front, and I spent 11 months as a prisoner of war in an "officer prison camp" in a former convent in the city of Monopoli in Apulia. We inmates, about 130 Austrian and Hungarian officers, made every effort to use our time as well as possible. Luckily our camp commander, Capitano, was a very kind

I volunteered for duty in the Austro-Hungarian army in 1913 and served in the Second Kaiserschützen regiment in Bozen, a "mountain infantry regiment".

With my fellow soldiers, (second from right and now a lieutenant), 1916. Soon after my entry into the Imperial Austrian Army, the First World War broke out, so that I had to move directly to the battlefield without seeing any of my relatives. I was wounded three times during the war, fortunately never very seriously, and I recovered rather rapidly. We all survived. At the end of the war in February 1918, my whole division was captured by the Italians and we were transported to a prison camp near Bari in Southern Italy. It was not until September 1919 that I was able to return to Vienna and see my parents after five years of absence.

and civilized man, a high school teacher from Trapani in Sicily. He permitted us to buy books of all kinds. and we all studied languages—Italian, French, and English—by studying grammar and vocabularies and by reading newspapers and novels. We became rather fluent in all these languages, and even learned a little Spanish. A few comrades who had spent several semesters at various universities started to give courses in mathematics and physics. With the help of Italian textbooks, I organized a course in general chemistry.

Note: Modestly, Dr. Mark did not dwell upon the hardships involved during the war. However, G. Allan Stahl has written a moving description of what transpired during Mark's military service:

On June 23, 1914, as Mark neared the end of his service, the events of Sarajevo changed everything. The Kaiserschutzen regiment was immediately reassigned to combat, not in the mountains, but on the plains of Galicia on the Russian front.

Ill-trained for defensive warfare on a massive front, [the] Austrians fell back before a coordinated Russian attack. In the ensuing two months they retreated nearly 300 miles. Corporal Mark, however, frequently stood out. In the bitter battles of Lvov and Przemysl, in which the regiment reportedly lost 85% of its men, Mark was decorated for bravery. Once, in a rear guard action, he single-handedly held 10 Russians at bay until his unit was safely withdrawn. His stamina kept him in good stead, just out of harm, until in September, 1914, he was shot in the right ankle.

Wounded, he was evacuated to Hungary, where it was ascertained that the bullet had broken a bone, and he would require additional hospitalization. He was then sent to Vienna.

Back in Vienna, Mark spent the Christmas of 1914 healing at home. But rest and relaxation for this soldier took an unusual form, for during this break he began his study of chemistry at the University of Vienna. He completed one of the eight required semesters of study, and then returned with his regiment to the front. This time they were sent to the more familiar but no longer hospitable mountains along the Austrian–Italian border.

During the next 3 years Austria and Italy fought [to] a stalemate, neither side able to win a decisive victory. Slowly, however, the tide began to rise in favor of the Italians. In those days, the desperate Austrians used the Kaiserschützen regiment as an attack force to regain lost terrain. On one such occasion Lieutenant Mark, now decorated for valor on numerous occasions, won one of Austria's highest awards for bravery, the Leopolds–Orden, at Zugna Torta.

In the summer of 1918, the Italians captured Zugna Torta, a peak, and threatened the city of Trento. Tossed into the conflict, Mark's regiment suffered greatly trying to recapture the heights. In a final and desperate drive, Mark led several hundred men over the ridge, and, thus, saved the

> city. During the same summer, he returned to the campus and completed another semester of study. The semester in which, as Schmid points out, his exceptionally high academic abilities were recognized.
>
> Mark returned to the front during the fall, but his sacrifices were in vain. The Austrian front collapsed in November, 1918, and Mark, now the most decorated company-grade officer in the Austrian Army, was imprisoned by the Italians.[1]

Thus, our time was filled with useful activities, and I have never since learned so much in so many different areas as in the Convento San Francisco in Monopoli. In September 1919 I learned that my father was ill. I bribed a prison guard and made my way back to Vienna, where I found my father somewhat better but still sick.

Chemistry Study in Vienna

1919–1921

> The concept of "free radicals" was not known in 1920—well, perhaps in politics, but not in chemistry.—Herman F. Mark[1]

When I returned to Vienna from Italy, I immediately took up my study of chemistry, which I had started while recovering from a wound during the war. By far the most attractive professor in chemistry at that time was Wilhelm Schlenk. He had come to Vienna a few years earlier from Munich. For those interested in "scientific genealogy", Schlenk was "descended" from the illustrious line: Liebig–Kekulé–[Adolf v.] Baeyer–Willstaetter–Schlenk.

Schlenk's field of research was the "trivalent" carbon compounds, molecules of great reactivity and of eminent theoretical interest. He was a researcher of great imagination and, at the same time, an inspiring teacher and educator. We all followed his lectures with great enthusiasm because he conveyed not only information, but also basic understanding and perspectives. He was an eminently skilled experimentalist, and his work with his hands in the laboratory was as exemplary as that with his head when he gave lectures or wrote papers.

My thesis project was the synthesis of pentaphenylethyl. I assumed that octaphenylpropane would be unstable and

Immediately after returning from the army, I enrolled at the University of Vienna as a student of chemistry under Professor Wilhelm Schlenk, who also supervised my thesis. Here in Schlenk's laboratory, I successfully completed a doctoral thesis on the scientifically interesting and challenging subject of the synthesis of pentaphenylethyl in July 1921, at the age of 26, with distinction.

Professor Wilhelm Schlenk (third from left, back row) with his students, 1920. I am second from the right, back row, and my close friend, Hans Endar, is at the far left.

obtained the desired compound by the reaction of triphenylmethylsodium with dichlorodiphenylmethane:[2]

$$2Ph_3CNa + Ph_2CCl_2 \longrightarrow [Ph_3C{-}C(Ph_2){-}CPh_3] + 2NaCl$$

$$\downarrow$$

$$Ph_3C\cdot + Ph_3C{-}\overset{\displaystyle Ph}{\underset{\displaystyle Ph}{\overset{|}{\underset{|}{C}}}}\cdot$$

After isolating the pentaphenylethyl, I recorded its spectrum and proved its structure by conversion to pentaphenylethane and pentaphenylchloroethane. At this point Professor Schlenk asked me for a sample of my compound and asked, "Have you made an elemental analysis of it?" When I said I had, he said, "Let us see how good an analytical chemist you are." He carried out the analysis with his own hands, using the clumsy equipment of that time–a procedure that took about 2 hours. When he was satisfied, he nodded and said, "I think you can start writing your thesis." This is the kind of teacher who makes you love your profession!

The Move to Berlin-Dahlem

1922–1926

In 1921 Schlenk was invited to the University of Berlin to succeed Emil Fischer, who had died in 1919, and he asked me to come with him. A year later Fritz Haber, director of the Kaiser Wilhelm Society, was organizing an institute of fiber research, and he asked Schlenk to recommend a "modern organic chemist" for this project. Schlenk very kindly recommended me, and so I came to join the most outstanding scientific institute of that time.

Early in 1922 I married Marie (Mimi) Schramek, a Viennese like myself. When we first moved to Berlin-Dahlem, where the institutes of the Kaiser Wilhelm Society were situated, we had to live quite modestly. My salary was small, and we lived in a one-room apartment that included the occasional use of the kitchen. In the morning I bicycled to the institute, where there was always interesting and exciting work to do. Mimi was an excellent housewife, a superb cook, and a skillful seamstress. She kept the conditions of our life perfectly comfortable. We loved each other dearly and used to go after dinner to a nearby movie or dance hall. I told Mimi what I was doing or trying to do, and she told me about her plans—a larger and newer apartment, children, visits to Vienna, and many other things.

Newlyweds—just before we left Vienna for Berlin, following our wedding in 1922.

After World War I, the German textile industry was severely suffering from the competition of English, Dutch, and French companies, which had easier access to the essential raw materials and processing equipment. It was, therefore, decided in 1920 to establish, within the framework of the Kaiser Wilhelm Society (renamed Max Planck Society after World War II), an institute for fiber research. This institute would study the fundamental structure of fibrous materials and its relationship to their technically significant properties.

R. O. Herzog, a prominent representative of fiber sciences and technology, was appointed director. He decided that he would try to make important contributions on the molecular structure of technical fibers with the aid of newly developed experimental methods such as X-ray diffraction and ultramicroscopy. He had assembled a group of scientists for this task under the leadership of Michael Polanyi, who started as a physician but, by that time, had become a physical chemist and was later to become a preeminent social and economic philosopher. Karl Weissenberg, a mathematician and theoretical physicist, would be our mentor and "father confessor" in all complicated derivations and computations. Rudolf Brill, a graduate student, and I were supposed to take care of the experimental part of the studies. We started by studying the crystal structures of a wide variety of substances, e.g., hexamethylene tetramine,[3] pentaerithritol,[4] urea,[5] triphenylmethane,[6] graphite,[7] oxalic acid,[8] carbon dioxide,[9] and ammonia.[10]

Later, the objects of our attention were the natural textile fibers (silk, cotton, hemp, and wool), all of which belonged to a large group of technically important materials that were generally classified as high-molecular-weight natural organic substances. (At that time, *hochmolekular* did not necessarily imply large molecules, but was frequently used for substances believed to consist of aggregates of small molecules.)

This way, one year after another went by. Difficult but lucky research efforts were joined by love and devotion. Then one day there knocked at our door, hesitantly at first but later more and more emphatically, a dear and welcomed guest—science. It came, was warmly greeted and carefully acknowledged, and it stayed.

Early Studies with Natural Fibers

Cellulose

A few interesting observations about the crystallinity of a cellulose film had been published and had instantly placed our group (Michael Polanyi, Erich Schmid, and me) in the forefront of useful X-ray research of solid materials. Now it was necessary to fortify this position by newer experiments and better interpretations. We received a contract, and we had suddenly climbed the first, most difficult, rung of the ladder to success. We immediately learned an important lesson: Concentrate on one problem; do not dilute your effort by working in several directions at the same time. Within a month I was made senior research associate at our institute, with twice the old salary and with a vastly increased budget for equipment and laboratory help. Gradually our work became known to our contemporaries, first in Germany (O. Goldschmidt) and later in England (J. D. Bernal and W. L. Bragg), France (L. De Broglie and J. J. Trillat), and Sweden (M. Siegbahn and Th. Svedberg). We were invited to international congresses. By 1925 we were accepted members of the community of fiber chemists and rheologists using X-rays as the main tool of their research.

In 1920 Hermann Staudinger had postulated (for rubber and for two synthetic resins—polystyrene and polyoxymethylene) the presence of long-chain molecules with very high molecular weights.[11] He did not include cellulose in his paper, but Polanyi,[12a, 12b] who had evaluated X-ray diagrams of ramie fibers made by Herzog and W. Jancke, arrived at the conclusion that the unit cell obtained from the diffraction pattern contained either segments of long-chain molecules or cyclic disaccharides. (Forty years later Polanyi wrote: "Unfortunately I lacked the chemical sense for eliminating the second alternative."[13])

A little later O. L. Sponsler and W. H. Dore (a botanist and a crystallographer) proposed on the basis of their X-ray data a cellulose chain structure that disagreed with unambiguous chemical evidence,[14] and this discrepancy made organic chemists very skeptical about the usefulness of crystallography. Eventually, after my move to the I. G. Farben laboratories, I reported with K. H. Meyer a cellulose structure consistent with both the crystallographic and chemical data.[28] Our ongoing work with organic fibers, metal wires, needle-shaped crystals, and whiskers was greatly facilitated by a simple geometrical formula that Polanyi had derived as soon as he became interested in X-ray diffraction.

In 1912 Max von Laue, the discoverer of X-ray crystallography, had given a general derivation for the selective reflectivity of X-rays through a three-dimensional lattice. W. H. and W. L. Bragg soon published a simple equation—Bragg's reflection law— that shows very clearly how X-rays are reflected from a stack of crystal planes. Polanyi, who was interested in the study of oriented fibers, derived a similar law for X-ray diffraction by a bundle of linear lattices.[12,15] This "layer line relation" was of great value for our studies and gave very direct information about the identity period of the investigated lattice. Even the study of one specific cellulosic fiber—native ramie or viscose rayon—provided a wealth of interesting information. It became qualitatively evident that these systems were built up of two phases: small crystalline domains embedded in an amorphous matrix. Swelling destroys the lateral order of the chains (i.e., crystallinity), which returns if the swelling agent is removed. For relaxed fibers, the orientation degree of the crystalline domains is reduced; under tension it remains the same or even increases.

Staudinger, in his classical elaborate studies on macromolecules, had repeatedly and correctly emphasized the fact that they produce solutions with high viscosity. He had even proposed a law—the famous Staudinger viscosity law—to use this viscosity for the estimation of molecular weight.[16]

We, in our fiber research, were interested in many other properties of our materials, namely strength, elasticity, water absorption, and abrasion resistance. Until this point, for solid organic compounds, interest was focused on melting point, solubility, color, surface activity, and many pharmaceutical and medical properties, but never on mechanical behavior.

We soon recognized that the solution viscosity of polymeric systems is not their most important property, but the enormous influence of polymer chain length on all physical and mechanical characteristics is. In fact, the astounding influence of native polymer structure on mechanical behavior was known for centuries in the hardness of ivory and ebony, the elasticity of kangaroo tails, the toughness of alligator skins, and the softness of cashmere and vicuna wool. Here, obviously, was a large field for new and useful research. Polanyi, who had been and still was working on the mechanical properties of metals, took the lead in following this new direction.

Thus, between our X-ray tubes and our textile machines, we treated our fibers by swelling, stretching, relaxing, and drying. We established changes in their crystalline–amorphous system and explored how the macromolecules react to changes in their physical and chemical environment. Two general principles emerged as controlling factors: orientation and crystallization. Together these factors can explain a variety of properties such as rigidity (tensile modulus), extensibility, elasticity, melting, softening, and swelling.

Whenever long-chain molecules are oriented by stretching or shearing, two different results are observed. If the molecules are irregularly structured (cellulose acetate, starch, wool, and collagen), their axes become parallel, and kinks or twists are straightened out. Then, as soon as the orienting force ceases to act, the system returns (sometimes slowly) to its original state. However, in the case of regularly structured chains (cellulose, silk, chitin, and rubber), after the axes become parallel a new phenomenon occurs: crystallization.

Our X-ray diagrams clearly showed that, as a result of the orientation, intermolecular forces produce a system of small "crystalline" (laterally ordered) domains that have a much firmer structure and do not disappear upon relaxation. We used to say that a kind of "molecular zipper" runs along the oriented chains and "reinforces" the system.

Rubber

In this context, a particularly dramatic observation was made by our colleague, J. R. Katz. While working in Copenhagen in the laboratory of Professor F. Hansen, Katz found that natural rubber, which is amorphous in the relaxed state, crystallizes on stretching. This was an entirely new phenomenon.[17] It suggested that rubber consists of long-chain molecules and led me (in collaboration with E. A. Hauser and P. Rosbaud) to study the crystal structure of rubber in parallel with our efforts on cellulose and silk. We showed that the repeating unit in rubber is

$$\left[-CH_2-C(CH_3)=CH-CH_2-\right]_n$$

rather than the *trans* form advocated by Staudinger, who chose to ignore our unambiguous evidence. As Linus Pauling remembered this period:

> Mark's X-ray work on fibrous macromolecular substances began in 1925 with his publication, together with Katz, of a paper on cellulose. He continued the work on cellulose with Meyer and Susich. In 1926 he and Hauser published a report of their studies of rubber. He had developed excellent ideas about the nature of rubber and the explanation of its extensibility and elasticity. I remember that when I visited him in Ludwigshafen in the summer of 1930 both he and I took pleasure in a demonstration that he showed me. He took a large piece of unvulcanized rubber, and stretched it to twice its length. When it was released, it contracted to its original length. He then stretched the rubber and held it under a cold-water faucet, so that it was cooled in the stream of water. On being released, it remained in the stretched

Our X-ray laboratory in Berlin-Dahlem, 1926. Drs. H. Kallmann, left, and Paul Rosbaud. It was a relatively crowded, small laboratory, where great care had to be taken not to disturb any of the photographic equipment.

form, which had crystallized. He discussed the part played by the increased entropy of the contracted form in the extensibility of rubber, much to my edification.[18]

Fiber Tensile Strength

Our new tendency to invoke molecular phenomena for the explanation of macroscopic behavior also led to the first attempt to arrive at an estimated upper limit of the tensile strength of a fiber.[19] The argument is as follows: It is known that the work required to break a covalent bond (carbon–carbon or carbon–oxygen) is on the order of 70 kcal/mol. Because we knew the exact structure of cellulose, we could calculate that the work necessary to tear a single chain would be about 5×10^{-12} erg. If the dependence of the force between two separated chain ends on their distance from each other is substituted by a step function in which the force decays to 0 for a separation of 2.5 Å, then the force necessary to break a chain is about 2×10^{-10} kg per chain or, from the structural model of cellulose, about 800 kg/mm^2. This is much larger than the largest experimental values then known, which were in the neighborhood of 100 kg/mm^2. Apparently, as in the case of metals and inorganic materials, the reasons for the discrepancy are flaws and imperfections that reduce the tensile strength far below its theoretical value. This somewhat crude estimate was later repeatedly refined, but the order of magnitude remained the same.

Superimposed on all these structural considerations was the influence of the molecular weight (i.e., the length of the individual chains). To arrive at experimental data we prepared viscose and acetate samples from wood pulps of different molecular weights by starting with a pulp of molecular weight around 200,000, corresponding to a degree of polymerization (DP) of anhydroglucose residues around 1200, and degrading it to DPs of about 1000, 800, 600, 400, 200, and 100. Cellulose acetates and cellulose xanthates were prepared from these samples, films were cast, and their tensile strength was measured in the air-conditioned state. Up to a "critical" DP (about 100), no strength at all is developed. The samples (fibers or films) are very brittle; thereafter strength increases roughly linearly with the DP until, after a downward bend, it reaches a plateau. From then on, it increases only slightly.

Clearly, this means that below a critical value DP_0 there is no strength at all, whereas above another critical range DP_w (the weight average DP), an additional increase of DP pays no more dividend on strength. We accepted this at the time as useful empirical information. Later, P. J. Flory showed that it had a sound theoretical basis.

In the end, the conclusions from our work on natural fiber formers were simple enough. They all consist of long-chain molecules with molecular weights well above 100,000. Some are regularly built and can crystallize; others are not. They all behave like organic substances with respect to stereochemistry, composition, and reactivity of functional groups (hydroxyl, carboxyl, amino, and others). Covalent bonding along the chains and van der Waals interaction between them made most of their properties understandable, at least in a good qualitative sense.

Quantum Physics and Albert Einstein

Truthfully, the results of our studies failed to impress the leading members of the scientific community in the Kaiser Wilhelm Institute, including Max von Laue, Fritz Haber, O. Hahn, Lise Meitner, James Franck, K. F. Bonhoeffer, and others who were preoccupied with radioactivity, atomic and molecular quantum phenomena, and catalysis. However, word was passed around that we had powerful high-voltage and high-vacuum installations and X-ray tubes of unusual intensity, monochromaticity, and durability. This equipment attracted occasional visitors from the other institutes to our laboratory and established some interest in our work.

Albert Einstein, probably the most celebrated and aggressive theoretical physicist of all times, was nevertheless always very much interested in experiments. When he visited our institute (about a dozen times altogether) he always wanted to know what we were doing with our X-rays and what our special problems were. We had, at that time, a few fellow physicists, S. N. Bose, P. P. Ewald, L. Szilard, and Eugene Wigner. During Einstein's visits, my wife, Mimi, came down with tea and sandwiches. I remember many of these "bull sessions" in which the great man told us stories about his life. On one occasion he suggested a new way to establish the existence of light quanta, but made a mistake in his calculations. Szilard, who was very sharp, noticed it and said, "But, Professor Einstein, what you are

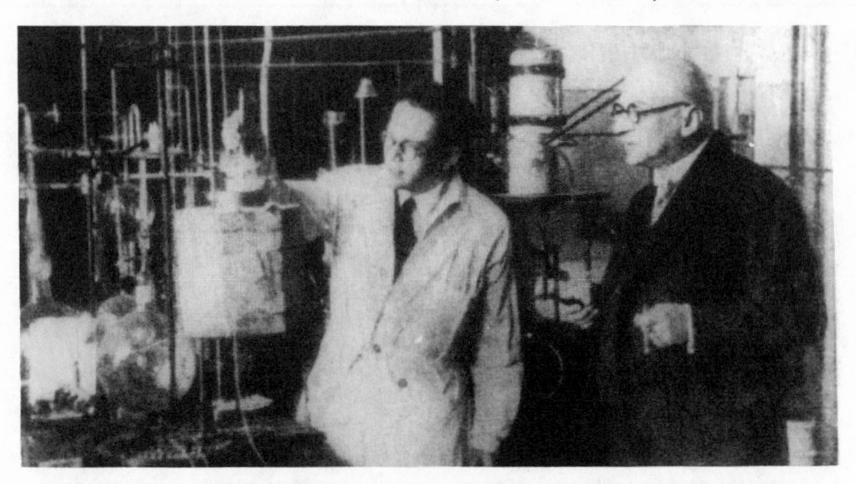

Dr. N. Bonhoeffer (left) demonstrates an experiment to Dr. Fritz Haber (right) at the Kaiser-Wilhelm Institute, ca. 1924. Haber was the Director of the Institute of Physical Chemistry in Berlin-Dahlem.

With Professor S. N. Bose (right) of Dacca, India, (now Bangladesh), who was one of many distinguished foreign visitors to Berlin-Dahlem, ca. 1925.

just proposing won't work." Einstein smiled and said, "Yes, I am not good at rapid calculations, but if you give me enough time, I am, in the end, usually right."

The breakthrough occurred when, one day in 1924, Einstein came and asked us to verify, if possible, the famous "Compton effect." This experiment involved the scattering of X-rays by free electrons. It was, at that time, the strongest support

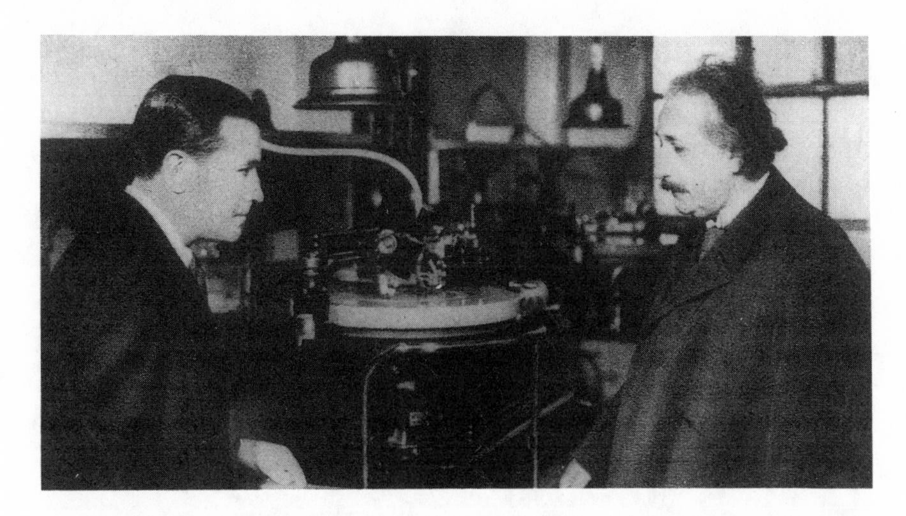

Showing our X-ray diffractometer to Professor Albert Einstein in Berlin-Dahlem, 1923.

of the light quantum theory, for which he had just received the Nobel prize in physics. A. H. Compton of Washington University in St. Louis had discovered this phenomenon in 1923, but W. Duane of Harvard had not been able to confirm it. Because we had exceptionally intense X-ray tubes, we might be in a position to make an important contribution toward resolving this burning question. We went to work immediately; within a few months we were able to confirm the existence of the wavelength shift observed by Compton and to contribute additional information on the nature of the Compton radiation.[35,36] This contact with an important problem of quantum physics brought several famous visitors to our institute in addition to Einstein (such as Max Born, Arnold Sommerfeld, and Hans Thirring). It also initiated further work on X-ray physics with P. P. Ewald, L. Szilard, E. Wigner, and H. Kallmann.

I. G. Farben, Ludwigshafen on the Rhine

1927–1932

> Once in Ludwigshafen the laboratory director, O. Seidl, told me, "Mark, you should go into business. If you are as successful in making money as you were in spending it on research we can already congratulate ourselves."[1]

Some time in the summer of 1926, *Geheimrat* Haber called me and asked me to visit him in his office. There he told me, "Dr. Mark, you have now, for 4 years, with the aid of X-ray diffraction, investigated many crystalline materials, elements (Zn, Sn, C), simple organic and inorganic compounds (NH_3, CH_4, BH_3), and finally natural fibers of all kinds. Your experiments have been well planned and executed; the results are of fundamental value and are well presented. You could now continue your research efforts along the same line and become a specialist in the X-ray investigation of solid substances. But you could also change the target of your activities and attempt to draw practical consequences from your new basic results. In a few days an old friend of mine, Kurt Meyer, director of I. G. Farbenindustrie, will come to visit me. He wants to talk with you about your future life: either to stay here and continue the style of your work or to move to a practical application of your results." I. G. Farbenindustrie was the largest

German chemical corporation at the time. My general answer was that if Professor Meyer would allow me to apply the present knowledge on fibers to the production of improved species and at the same time continue my fundamental studies, I would be very glad to accept any of his proposals.

When he interviewed with Meyer in 1926, Mark outlined a typically thorough program. He proposed a team of organic and physical chemists and physicists who would evaluate the influence of structure on properties such as rigidity, elasticity, melting point, and water absorption. Work, he proposed, would shift toward new material development and into the manufacturing facilities to evaluate the effects of processing on structure as their knowledge base expanded. "All of this," Mark told Meyer, "can be accomplished if I have the necessary apparatus, equipment, and a few able physicists, physical chemists, and, of course, a good organic chemist."

Meyer approved and hired Mark. In doing so he began simultaneously a very close, life-long friendship and a professional relationship which would help bring I. G. Farben and polymer science to new heights.

In October 1928, a very interesting conference on quantum mechanics was held in the Italian city of Como. Here I am with Professor Kurt H. Meyer at the shore of Lake Como.

> . . . Mark's contributions while at the I. G. were not limited to the emerging field of polymer science. In those five years, he also took part in studies of X-ray optics and continued his study of the X-ray structure of metals and metal salts. Other seemingly unrelated papers were published on the width of X-ray emission lines,[20] Schlenk isomerism,[21] the structure of aromatic compounds,[22] and a special "hobby", the optical Stark Effect.[23,24] Regarding this latter work, Mark relates that his supervisors tolerated the research, commenting that "as long as they are doing something decent and important" it was okay, "as sport doesn't cost much money".[1]

In the middle of December 1926, Mimi and I spent Christmas vacation with our families in Vienna. By the end of the month we had moved to Mannheim, where we lived for the next 6 years. This old, middle-sized city at the confluence of the Rhine and the Neckar rivers was at that time a busy center of agriculture, industry, and ship-to-railroad traffic. The valley of the middle Rhine (at that point about 20 miles wide) is an extremely fertile, highly developed area with excellent crops ranging from corn and other garden vegetables, to quality fruit, to wine. There were thriving industries ranging from paper and furniture to chemicals of all types, such as sulfuric acid, caustic soda, ammonia, dyestuffs, and pharmaceuticals. The Rhineport of Mannheim was a busy exchange center for the export of local products and the import of bulk materials such as coal, ore, and wood. The society was very cultured, with theater, music, architecture, and fashion. Mimi, my wife, with her Viennese charm, was soon very content and happy in the midst of it all. We had delayed having children in Berlin. Now that our situation seemed secure, we decided to have a family. Hans Michael was born in 1929, and Peter Herman was born in 1931.

The Move to the Industrial Research Laboratory

For me, the change of activity was somewhat more complex and not without problems. At the Fiber Research Institute I had mainly worked with *things*: X-ray tubes, cameras, high-voltage equipment, and photographic chemicals. I relied on my own hands and personal experience, and I operated on my direct responsibility. Now I had to work with *people*: advise and supervise collaborators or laboratory helpers, and explain to them what to do and why. Later, when they came back with their results, I had to check on quality and reliability without complete familiarity with all the details. It was for me the first step on the ladder of administrative activities. The work involved much teaching and counseling on my part. Most of the issues were quite familiar to me, whereas for most of my co-workers the same issues were new fields of experience that had to be explained to them.

Fortunately, Meyer had consented that I bring two of my closer co-workers from Berlin with me. G. von Susich and E. Valko were of great value in "transferring" the new art of fiber and polymer science from Berlin to Ludwigshafen. Meyer also gave me valuable and needed personal advice, saying for instance:

My collaborators at the main laboratory at I. G. Farben display the electron diffraction instrument, 1928, on the occasion of a visit by George de Hevesy (seated). Standing (left to right) M. Dunkel, J. Hengstenberg, K. Wolf, G. von Susich, E. Valko, R. Wierl, and C. Wolff. I am on the far right.

> Make your group large enough and of high academic caliber. I want you to maintain close and friendly relations with universities in our neighborhood; for that you must have first-class scientists. Because our fiber- and resin-producing personnel ought to know more about the processing and handling of their materials, you will need energetic and inventive chemical engineers.

Close contact with high-level scientific circles was an invaluable heritage from Berlin-Dahlem. I also knew that the most important ingredient for progress was a clear-cut superiority in one of the leading experimental techniques. We had it in Dahlem in the X-ray field, and we rebuilt it in Ludwigshafen by adding electron diffraction and spectroscopy (optical and IR). Electron diffraction of gases was only of scientific interest and could not possibly contribute to the business of I. G. Farben. It was a remarkable example of their liberal attitude that I. G. Farben allowed me to pursue this field. Incidentally, when Linus Paul-

Professor J. Prins (seated, center) of the University of Leyden, on the occasion of his visit to our laboratory in Ludwigshafen. I am listening to Prins' comments to R. Wierl (seated, right). Behind him (left to right): J. Hengstenberg, R. Brill, G. von Susich, E. Valko, and K. Wolf.

ing visited us in 1930, he was delighted to learn about this new method for studying molecular structure. As Dr. Pauling recalled the occasion:

> In 1930, when I visited Herman Mark in Ludwigshafen, I learned that he and his young associate R. Wierl, had constructed an apparatus for scattering a beam of electrons from gas molecules and had determined the interatomic distances in carbon tetrachloride and a number of other molecules by analysis of the diffraction pattern (*Die Naturwissenschaften 18, 205, 778, 1930,* and later papers by Wierl in *Phys. Zeit.* and *Ann. Phys.* in 1931 and 1932). The equations describing the diffraction pattern produced by a wave (X-rays or electrons) scattered by a molecule had been derived independently by P. Ehrenfest and P. Debye in 1915. The electron-diffraction pattern from a molecule such as carbon tetrachloride showed a series of concentric rings, with different intensities, and with the radii of the rings and their intensities depending upon the interatomic distances and the scattering power of the atoms in the molecule. I was overwhelmed by my immediate realization of the significance of this discovery.
>
> . . . As the impact of the significance of this discovery burst upon me, I could not contain my enthusiasm, which I expressed to Mark—my feeling that it should be possible in a rather short time, perhaps ten years, to obtain a great amount of information about bond lengths and bond angles in many different molecules. I asked Mark if he and Wierl were planning to continue with such a program, and he said they were not. He added that if I were interested in building an electron-diffraction apparatus he would be glad to help, and in fact he gave me the plans of their apparatus. On my return to Pasadena in September I talked with a new graduate student in the California Institute of Technology, Lawrence Brockway, about this project, and he agreed to undertake the construction of the apparatus (with the help and advice of my colleague, Professor Richard M. Badger). During the following 25 years the structures of molecules of 225 different substances were determined by the electron-diffraction method in the California Institute of Technology, through the efforts of 56 graduate students and post-doctoral fellows. These studies led to the discovery of several valuable principles of structural chem-

> istry. I continue to have a feeling of gratitude to Herman Mark for his discovery of this important technique and for his generosity to me in connection with it.[18]

Professor Meyer had excellent contacts with the University of Munich, and after 1 year I had assembled a team of physicists—R. Wierl and K. Wolf from Munich, and J. Hengstenberg from Freiburg—who continued, with G. von Susich, our structural studies of polymeric systems.

Academic Contact

In the neighborhood of Mannheim, along the Rhine, there were several well-known universities, (including Darmstadt with E. Berl, Heidelberg with K. Freudenberg, and Freiburg with H. Staudinger). All three of these professors had been working on natural high-molecular-weight substances for many years. Professor Berl, a specialist in the technology of cellulose and its derivatives since 1906, helped us to transfer our basic knowledge of these materials into practical applications in the fields of fibers, film, and plastics. Professor Freudenberg in Heidelberg came from the famous group of Emil Fischer in Berlin.

Fischer had, at the beginning of the century, already demonstrated the existence of polypeptide chains in proteins.[25] Yet Fischer doubted that such chains could exceed a molecular weight of 4000,[26] and his prestige was one of the factors that made it more difficult to sell the macromolecular concept.[27] Freudenberg himself had carried out in 1921 a careful kinetic study of cellulose degradation and had concluded that a long-chain structure of this substance was in full agreement with his experimental findings. Professor Staudinger in Freiburg had already postulated long-chain structures for rubber, polystyrene, and polyoxymethylene.[11] In the late 1920s Staudinger's institute was a famous center of macromolecular science. Thus we were surrounded by schools of higher learning that were all generously aided by industry, and particularly by I. G. Farben. Karl

The Wilhelm Exner Medal was presented to me in Vienna, 1934. Professor Karl Bosch was on hand for the occasion because of his interest in my work during that period. Left to right: me; H. Thirring, the Italian ambassador (who represented the legendary Guglielmo Marconi); Dr. Bosch; W. J. Mueller; and H. Fischer.

Bosch, president of the corporation, once commented, "We give them money to educate their students better. They do that and we hire the better students, with whom we manage to make even more money." A very encouraging remark for the support of academic institutions by industry!

Further Studies with Cellulose

What was obviously expected from our newly founded laboratory was to use the new basic understandings and insights to arrive at desirable and necessary practical results. To do that, we had to expand the horizon of our efforts, because by and large, at the Fiber Research Institute in Berlin-Dahlem we had focused our attention on structural analysis and fundamental behavior. Now we would have to expand into synthesis and application. For such an expansion of our capabilities, we needed imaginative organic chemists and inventive chemical engineers. For synthesis we had H. Hopf, K. Guenther, and W. Ebel; for applications we hired S. Biedenkopff, M. Dunkel, and R. Brill.

Fiber Structure

Organic synthesis was sorely needed in the chemical preparation of cellulose derivatives, mainly the xanthates and the acetates. Structural analysis and the fundamental procedure of our existing fiber production had to be considerably expanded. The best applications for fibers were entirely dependent on the numerical values of such properties as tensile modulus, tensile strength, and elongation to break.

We knew from the work in Dahlem that cellulose fibers contain certain crystalline domains; now we had to find their exact sizes and shapes. We found that these crystallites in a viscose rayon fiber have a rodlike shape, about 20–100 nm long and 5–10 nm wide. Its various derivatives (nitrates, acetates, xanthates, and others) led to different crystalline modifications. The arrangement of the glucose units along the chain was always the same because it depends on ordinary covalent bonds of about 1.4 Å. However, lateral packing is regulated by several types of intermolecular forces between polar and readily polarizable groups. (The concept of hydrogen bonding did not yet exist.)

A series of X-ray diagrams taken under carefully controlled conditions showed that extended rubber molecules assume different conformations, depending on the conditions of the deformation, such as the rate of stretching and the temperature of the experiment. This variation shows that many intermediate states of orientation exist between the random coil and the fully extended chain. Evidently, these intermediate states of orientation are the reason for the hysteresis phenomenon, which, in turn, is important for the heat developed during periodic extension and contraction of elastomers. We encountered a surprising number of phenomena during these years of fundamental polymer research, which essentially represented a continuation of the "Dahlem working style". Most important was the work on the crystal structure of cellulose,[28] silk fibroin,[29] and Heves rubber,[30] as well as electron diffraction by gases.[31]

Solution Spinning

Meyer's original assignment, as already mentioned, included close cooperation with existing production lines and with future potential developments. One obvious call for such a practical activity came from the cellulose division of I. G. Farben, which made fibers from cellulose xanthate and cellulose acetate, together with sheets from cellulose nitrate. With the chemists of the fiber division (F. Gajewski, P. Schlack, and P. Kleine), we developed the first fragmentary theory of "solution" spinning.

> It is interesting to note that in their first paper on cellulose (1928) Meyer and Mark proposed a structural unit cell model which is classic and accepted, for the largest part, even

> today. They proposed a cellulose crystallite in which all atoms were bound to each other in very long chains by primary valence forces, and the chains in turn aggregated in larger combinations by secondary forces. The simplicity of this concept is misleading. It represented a compromise between the association theory of molecular interaction and macromolecules, while actually embracing the latter.[1]

Unlike many famous scientists, our engineer friends had no difficulty in accepting the concept of flexible long-chain molecules. The spinning solution, obviously, should be a strongly entangled system in the sol state. Upon emergence from the spinnerette and entry into the spinning bath, it is converted into a gel by osmotic or thermal removal of the solvent. In this firmer and elastomeric state, it is stretched to produce the desired degree of orientation and, eventually, the crystallization necessary to arrive at the ultimate properties of the resulting fiber. Clearly, there were many degrees of freedom in this operation, and we expected that it should not be too difficult to adapt it to a wide range of useful textile characteristics.

On top of this relatively obvious sequence of events, we soon felt the urgent necessity for a much closer look at the intermediate systems and their quantitative control. Here the experiences of the years in Dahlem paid off.

First we dealt with the spinning solution. What was the best viscosity for the spinning process? Once this viscosity was established for a given spinning machine, say between 150 and 160 poises, was it better to get this "optimal" viscosity through a low concentration of higher-molecular-weight material or through a high concentration of a lower-molecular-weight species? Obviously, the engineers favored the second choice because it gave them a higher material yield per unit time. But maybe solutions of longer chains could be spun faster, the individual molecules might be better oriented in the faster flow; as a result the tenacity might be higher.

Good spinning was clearly the result of optimizing many parameters: DP and DP distribution of the cellulose, degree of substitution (acetylation or xanthation), concentration of the spinning solution, length and diameter of the spinnerette, spinning speed, and composition of the regenerating bath. It was always a jungle of operating conditions, and everything was empirical and delicate.

Most of the engineers in our fiber plants were classical organic or physical chemists. All were excellently trained in their own disciplines, but they had never heard the word "polymer" and were never faced with long, flexible, and randomly entangled chains. But it did not take long. They started to like our more fundamental approach, which gave them a vivid picture of what was going on at the molecular level and helped them to think about new ways to avoid difficulties and to improve their processes and products. Once they were on their own, the process supervisors and product development engineers generated additional favorable steps and invented new gadgets. The result was a modernized plant and a superior line of products.

Cellulose Acetate

One of our most attractive fibers was cellulose acetate; it was at that time and still is now—in spite of nylon—a top-level product. We produced it in several grades. The highest grade was a very fine (1–1.5-denier) continous filament of luxurious silklike luster and softness, together with a full hand and drape. It could be readily dyed, woven, and finished. Cellulose acetate was the only existing "replacement" for silk, which was so rare and expensive in Germany that, for all practical purposes, it did not exist on the marketplace.

Cellulose acetate, however, had one serious deficiency. Its tensile strength was low, only 2.5–3.0 grams per denier (g/den), whereas that of silk was above 4.5 g/den. In discussing this problem with the development engineers of our acetate plant in Berlin-Lichtenberg, we suggested that they carry out a few tests with a higher-molecular-weight species. Of course the viscosity of the spinning solution went up, the rate of its delivery decreased, and the filament yield decreased. However, the quality of the new HS (high-strength) fiber was so good that its production on a larger scale was authorized. During the designing of the plant our engineers introduced several changes: better filtration, higher spinning temperature, and the use of a different solvent, so that, in the end, the yield of the high-strength acetate fiber was as high as that of our standard product, with a tensile strength close to 4.2 g/den.

Studies with Rubber and Synthetic Polymers

We were also very much involved in the study of rubber, which had been shown by X-ray analysis to consist of long-chain molecules. As long as these chains remain chemically independent, the material is soluble. After mastication the bulk material is a thermoplastic mass, which means that it may be repeatedly softened by heat and may be readily molded or extruded.

About 100 years ago Charles Goodyear found that sulfur reacts with rubber in such a way that its thermoplastic flow is reduced or even eliminated and that a reversibly elastic solid mass is formed: "vulcanized" rubber. In Dahlem we carried out a close X-ray study of native rubber and several specimens with different degrees of vulcanization. We found that the sulfur reacts with the double bonds of the individual rubber chains and links them together by covalent C–S–C bonds. Instead of a mass of independent linear macromolecules that slide along each other, vulcanization produces a three-dimensional cross-linked network. The system that was originally plastic now becomes reversibly elastic.

Butadiene

Next to our laboratory Otto Schmidt worked on the sodium-catalyzed polymerization of butadiene, known as BUNA, i.e.,

J. R. Katz (seated), the discoverer of rubber crystallinity, examines an early X-ray photograph of stretched rubber at I. G. Farben, 1929. From left to right, G. von Susich, J. Hengstenberg, R. Wierl, K. Wolf, E. Valko, Ms. Carst, and me.

butadiene polymerized with natrium (natrium being the German word for sodium). Clearly, a three-dimensional network of polybutadiene chains had already been cross-linked during polymerization and, therefore, needed no subsequent vulcanization. Schmidt used butadiene in his work instead of the equally reactive isoprene because it could be conveniently obtained from oil or coal in the adjacent plant at Oppau.

After 3 years of organic chemistry at the universities in Vienna and Berlin and 5 years of polymer science in Dahlem, it was not difficult for me to realize that BUNA was a cross-linked polybutadiene. But why was it already cross-linked during the polymerization process, without any vulcanization? We knew that there are two types of polymerization catalysts, typified by benzoyl peroxide and metallic sodium. Today we know that peroxides initiate polymerization by thermal decomposition into free radicals and that sodium functions by an ionic mechanism, but these concepts were unknown in the 1920s. We found that, in the rubber plant, enzymatic catalysis is involved, and we concluded, incorrectly, that the catalyzed process is oxidation. This conclusion led us to the use of peroxides. Anyway, we found that peroxide-catalyzed polymerization leads to independent chain molecules, whereas sodium-catalyzed polymerization leads to cross-linked networks.

We had to find a way to polymerize butadiene with free-radical catalysts. Butadiene alone did not respond to the catalysts, but we had other monomers (styrene, acrylonitrile, and vinyl esters) that were very responsive to peroxide-catalyzed polymerization. Therefore, it was probable that butadiene mixed with such a monomer could be polymerized by peroxides. A series of systematic tests soon established that styrene and acrylonitrile were suitable comonomers. As expected, the new copolymers (BUNA-S and BUNA-N) consisted almost entirely of independent chains with very little cross-linking, and were, therefore, thermoplastic and easily processable. Much of this work was carried out with the very successful help of our colleagues at the I. G. Farben plant in Leverkusen: W. Bock, H. Lecher, and E. Tschunkur.

After elaborate tests in the laboratory and pilot plants and after road tests, these two synthetic rubbers, the first representatives of this species, together with Du Pont's neoprene, were made available to the automobile industry in the mid-1930s.

Copolymerization

The successful method of polymerizing two (or more) monomers in conjunction led to the new general concept of copolymerization. Hans Fikentscher carried out very careful and informative basic experiments, found that certain monomer pairs (e.g., butadiene and styrene) "like" to copolymerize, whereas others (e.g., styrene and vinyl acetate) do not. He divided all the monomers available at that time (about 30) into "compatible" and "incompatible", and this classification proved very useful in planning new experiments. However, the final clarification of the copolymerization problem had to wait until 1944, when T. Alfrey and G. Goldfinger,[32] as well as F. R. Mayo and F. M. Lewis,[33] derived the copolymerization equations for monomer pairs. Later Alfrey and C. F. Price explained the molecular origin of the copolymerization behavior and, for all practical purposes, closed the chapter on this part of polymer science.

For the experimental study of relatively rapid processes, such as the "mercerization" of cellulose or the extension and contraction of rubber, our physicists had developed high-intensity electron and X-ray beams. Once these beams were available, they were also used for other scientific investigations. R. Wierl used strong electron beams to study gases containing small and simple molecules (CCl_4, $CHCl_3$, C_2H_6, and others) to establish precisely their interatomic distances. P. P. Ewald and W. Ehrenberg applied our strong X-ray beams to develop the "double-crystal X-ray spectrophotometer" for the fine structure of X-ray lines. Although these activities were not of practical use, they were of considerable fundamental interest. Thus we gained a good reputation for purely scientific interests and exceptional experimental skill.

Our fiber plants all over Germany had been very satisfied with the new views and suggestions they had received from our polymer group. Now the BUNAs, with the very important contributions of our colleagues at the I. G. Farben plant in Leverkusen, had been launched as promising replacements for natural rubber, which had to be imported from overseas. In addition, they were very useful elastomers in their own right.

Polystyrene

There was still a large new field of polymers to be conquered: the synthetic thermoplastics, which could be molded, extruded, stamped, and thermoformed. Sure, we did have cellulose nitrate and acetate, but the nitrate is dangerously flammable, and the acetate is sensitive to acidic degradation, and the manufacturing processes were closely guarded by the Celanese Companies in England and France. We needed a new molding resin, entirely synthetic, not based on cellulose but on coal and oil.

In the late 1920s E. Dorrer, M. Dunkel, E. Roell, C. Wulff, and I discussed possible activity in the resin field. Somebody mentioned "polystyrene", and the word acted like a spark. The two base materials, benzene and ethylene, were readily available; the monomer synthesis falls into the domain of well-known high-temperature condensation reactions; and the polymerization had been carried out in the institutes of H. Staudinger and

F. Stobbe as well as, on a somewhat larger scale, in our own laboratory. Yet polystyrene was still a laboratory curiosity, and considerable process and product development would be necessary to make it a commercial material. It would probably be used in the same field as cellulose acetate in the form of a molding resin.

One of our customers, the Eckert and Ziegler Company in Cologne, was just developing a new process—injection molding—for which they wanted a material with a very little moisture uptake and less sensitive to acids. It was, therefore, decided that S. Biedenkopff, the head of our own engineering group, should work out the design for a 1-ton-per-day monomer unit and for a corresponding polymerization plant. During this period (4–6 weeks) we would speed up our laboratory work and produce enough of the material for practical testing by Eckert and Ziegler. Two months later, word came from Cologne that they liked the new material very much, but that several of its processing properties would have to be improved. After another year a somewhat larger unit went on-stream and a new resin (Trolitul) was turned over to one of our commercial production departments. It was a big hit.

Some Digressions

During those years we frequently had to confer with our colleagues at I. G. Farben-Hoechst and -Leverkusen. We took the train from Mannheim to Frankfurt, where the meetings took place. At the station in Mannheim, each of us bought a newspaper to read on the trip. Most of my colleagues took the *Frankfurter Allgemeine Zeitung* or the *Mannheimer Nachrichten*, but I always bought the *Voelkischer Beobachter*, the newspaper of the Nazi party. Frequently my colleagues asked me why I read this miserable paper. I used to answer, "If I want to know what is happening in Germany today, I shall read your newspapers, but I want to learn what will happen in Germany 4 or 5 years from now. Therefore, I read and believe the *Voelkischer Beobachter*." Unfortunately, I was right.

A chemist who does practical work in the laboratory produces new substances and methods to prepare them, but he also produces notes. They are condensed into weekly and monthly progress reports that grow in volume at all levels: for the bench chemist, for the supervisor, and most of all for the laboratory director. Some of these notes are used to formulate patent applications, and others are published in scientific journals. The bulk piles up and, in the course of time, leads to books that eventually will serve as a record of results achieved and will integrate these achievements into the existing state of the art. I wrote and published three books during my 10 years in Dahlem and Ludwigshafen: *The Use of X-rays in Chemistry and Technology*, *The Structure of the Natural Polymers*, (with K. H. Meyer), and *Physics and Chemistry of Cellulose*.

Confirming the Existence of Covalent Long-Chain Molecules

Since I started working on cellulose, silk, wool, and rubber in 1922 in Berlin, the scientific world has seen profound differences of opinion concerning the "true" molecular structure of the high-molecular-weight natural organic systems: cellulose, proteins, starch, and rubber.

In 1920 Hermann Staudinger, then a professor at the Eidgenössische Technische Hochschule in Zürich, had postulated that rubber and two synthetic resins (polyformaldehyde and polystyrene) consisted of very long chains that could have molecular weights in the hundred thousands and were held together by normal covalent bonds. Staudinger's assertion that covalently linked long-chain molecules exist rested on three types of evidence:

1. His students isolated a long series of formaldehyde oligomers and showed that the properties of polyoxymethylene (density, melting point, and X-ray diffraction) were those expected by extrapolation of the oligomer properties.
2. He showed that chemical modification of polymer side chains does not change their colloidal properties, although this should lead to a change in the aggregation of small molecules.

3. Increasingly, he based his arguments on solution viscosity, as interpreted by the viscosity law

$$\frac{\eta - \eta_0}{\eta_0} = K_m c M$$

where η and η_0 are the viscosities of the polymer solution and the solvent, respectively; c is the polymer concentration; and M is the molecular weight. The crucial K_m constant had to be "calibrated" by making viscosity tests with polymers, the molecular weight of which had been determined by other established direct methods such as vapor pressure reduction or osmotic pressure. Both methods work well in the low-molecular-weight range (up to 1000 or 1500) but become extremely unreliable in the higher ranges. Staudinger determined his K_m constants for various polymers (rubber, polystyrene, polyoxymethylene, cellulose acetate, and others) in the low-molecular-weight range and then extrapolated the relationship between the molecular weight and the solution viscosity of polymers to the domain of much higher molecular weights.

Many prominent organic chemists of these years (P. Karrer, H. Wieland, M. Bergmann, K. Hess, and others) doubted that this extrapolation was a scientific proof for the existence of macromolecules with molecular weights in the hundred thousands and higher. In their opinion, the characteristic properties of all these materials—solution viscosity, gel formation, strength, and high softening—could readily be explained by their colloidal character (i.e., by the aggregation of small normal molecules with the aid of strong intermolecular van der Waals forces).

On the basis of our own direct experience with many of these materials with respect to thermal stability, insolubility, mechanical strength, and elasticity, K. H. Meyer and I could not believe that the known intermolecular forces between small units could explain the enormous range of their properties. We therefore favored the existence of long-chain molecules. On the

During my stay at the main laboratory of the I. G. Farben, we had many distinguished and famous visitors. From the left are Professors Hückel, P. Debye, myself, K. H. Meyer, and P. Scherrer.

other hand, we could not accept the solution viscosity approach of Staudinger as a scientifically valid proof for the existence of such long chains. Hence, right from the beginning, we insisted that other dependable quantitative methods should be introduced to resolve the controversy, and we decided on the systematic use of X-rays. In fact, the data obtained in Dahlem and Ludwigshafen on cellulose, starch, chitin, silk, and rubber (and at the same time by W. T. Astbury in England on hair and wool) left no doubt of the existence of covalently bonded macromolecules.

> In an extensive paper, they carefully developed the idea of cellulose chains consisting of so-called "primary valence chains." They further proposed that the primary valence chains were aggregated by molecular forces such as hydrogen bonding and van der Waal's forces. Their model, which became a standard, combined the important features of both concepts by proposing micelles of long, not short, molecules. The physical properties of cellulose were attributed to these forces, for example, tensile strength to the primary valence bonds and insolubility to the secondary forces.[1]

Shortly afterward, Staudinger formally criticized our work. In a letter to me written October 31, 1928, Staudinger wrote

> Finally, I also have written a few papers opposing the views of K. H. Meyer. I do not agree with him on two points. First, in my opinion, the statements of K. H. Meyer do not represent anything new, but rather coincide in general with the opinions I have advocated for years and established experimentally. Second, I do not believe that the introduction of "primary valence chains", instead of macromolecules, solves any problem.[1]

I responded to Staudinger on November 2, 1928. Among other things I wrote

> I am sorry to see from your letter that you feel annoyed by the statements of Professor Meyer. I am convinced that it was the last thing Professor Meyer intended to do. I have, in our joint research and especially in my Hamburg lecture, emphasized the importance of your beautiful work. Introduction of the term, primary valence chain, is very purposeful since it refers to structures which are not identical . . . rather average lengths. If we add this fact to your macromolecules, then both concepts become identical.
>
> Because of this I prefer not to emphasize the differences . . . we mean the same thing. I believe that we should advocate this point of view together and not emphasize some slight differences in perceptions. The high polymer camp could easily make the mistake very well known in politics: because of small differences in neighboring opinions, a large point of view often does not get enough attention.[1]

Even though our own work confirmed and quantitatively strengthened Staudinger's basic concept, we could not agree with him on another issue of similar fundamental character. Staudinger repeatedly and unreservedly stated that macromolecules are long, thin rigid rods. In fact, his viscosity law was based on this concept, although the theory of this "law" was incorrect, even for rigid rods. For K. H. Meyer, H. Hopf, and me, the rigid rod concept was in glaring contrast to the well-known and firmly established phenomenon of rotation around covalent bonds. It was obvious that long chains would be subject to hindered rotations and would be intrinsically flexible. A very detailed and remarkably correct account of this scientific controversy has been given by Klaus Priesner in his book, *H. Staudinger, H. Mark and K. H. Meyer.* Interesting accounts may also be found in *Polymers: The Origins and Growth of a Science* by Herbert Morawetz and *History of Synthetic Fibers* by Herman Klare.

The Looming Nazi Threat

Thus in the early 1930s, under Meyer's continuous benevolent guidance, our group (by then about 30 scientists and engineers) had maintained and intensified top-level scientific contact with the academic world at home and abroad. Prominent persons (P. Debye, R. O. Herzog, R. Mulliken, L. Pauling, E. K. Rideal, P. Scherrer, N. W. Semenov, T. Svedberg, J. J. Trillat, and others) visited our laboratories, gave lectures, and attended seminars. Some of our board members liked these contacts because they amounted to an unsolicited, neutral, and impressive proof that the company acknowledged the importance of scientific research in industry and was ready to spend considerable funds on it. Other leading persons of our plant management liked even more the fact that three of our main production lines had already profited noticeably from the use of polymer science: the fibers by the appearance of our new "Aceta" type, the rubbers by the appearance on the scene of the first synthetic elastomers, and the plastics by "Trolitul" (polystyrene), a completely new and very attractive thermoplastic. Apparently nothing is more successful than success.

Then, out of the blue sky, the bolt fell. One day in the early summer of 1932, I received a phone call from F. Gaus, vice president and managing director of our plant. "Could you be so kind as to come to my office tomorrow afternoon?" I went, and after a friendly welcome, he said, "As far as I and all my colleagues on the board can see, Hitler will take over the political

In December, 1960, a symposium was held at Polytechnic Institute in honor of my 20th year at the Institute. Left to right: Peter Debye, 1936 Nobel Laureate in Chemistry; me; Linus Pauling, 1956 Nobel Laureate in Chemistry; and W. O. Baker, vice president for research, Bell Telephone Laboratories admiring a partially completed mural on the history of science at Polytechnic that was being constructed in the entry hall.

power in Germany within the near future. His principles and methods are only too well known. You are a foreigner, and I understand your father is, or was, Jewish. We hope that, even under the Nazi regime, this will not be a reason for dismissal, but it is quite sure that any promotion or advancement will not be possible for you. I suggest, therefore, that you look for another job—in academia, industry, or government—outside of Germany."

These were simple and clear statements, and they were true. Always the experienced leader and perfect gentleman, Dr. Gaus continued: "We have given you much freedom for your work and you have succeeded in establishing yourself as a scientist of high reputation here and abroad. We do not let you go because of poor performance but for reasons beyond your and our control; therefore, we shall pay your present salary until the end of our existing contract. If you need any additional help in moving to another place, please let me know. And now, good luck!"

After expressing my deep gratitude to Dr. Gaus for his candid and benevolent explanation of our future relations, I went home and discussed it with my wife. She agreed that we should leave Germany as soon as possible, because, in spite of all the attractive and satisfying conditions, it became clear that there existed in this country a progressive tendency for polarization, not politically in "right and left", nor socially in rich and poor, but emotionally in rational and fanatic. Most disturbing, this "people's movement" did not come from the people. It was started and carried on by a few frustrated and disgruntled members of the "civilized" part of the society.

Without delay, I started to look for a new job. After several unsuccessful approaches, a very interesting opportunity opened up, namely the position as a professor of physical chemistry at the University of Vienna. I accepted the offer and, late in the summer of 1932, we traveled to Austria. This time we had two small boys (Hans, 3 years old, and Peter, 1 year old) and an automobile.

Mimi and our sons, Hans and Peter, in 1936.

University of Vienna

1932–1938

When we arrived in Vienna in mid-September 1932, our contacts fit into three main groups. One was, of course, the family. My father had died in 1927, but my mother, brother, and sister were all there, together with innumerable uncles, aunts, and cousins. Mimi's entire family was also there. Many visits had to be made because, after all, we had been away for 10 years—a long time for all of us in our twenties and thirties.

The second group consisted of friends and comrades who had studied with me before the war and served with me in the army from 1914 to 1918. They were all my age or a little younger and yet had gone through hard times and trying experiences—battles, injuries, war prison, diseases, and loss of relatives. The most prominent of these friends was Engelbert Dollfus, who went through the war in my batallion, later became secretary of agriculture in the Austrian government, and in 1930 became chancellor of the Austrian Republic. He tried heroically to keep his country independent of the Nazis and the Socialists, but was assassinated during an abortive Nazi revolt in 1934. This group also contained many school friends who had attained important positions in life: G. Riehl as a surgeon; F. Indra as a lawyer; and H. Lauda, M. Gunthof, and E. Schaefer as leaders in the chemical, textile, and rubber industries, respectively.

Members of my third group of contacts were represented by E. Linhart (paper), J. Pollak (chemistry), and N. Kreidl (plastics). They all played an important role during my next 6 years in Vienna because they knew of my activities at I. G. Farben and sponsored projects of interest to them at my institute or even gave me outright support for fundamental research.

Academic Duties

My first duty at the university was to give the "main course" in physical chemistry, which had been given for many years by Rudolf Wegscheider, a prominent classical scholar of thermodynamics, whom I succeeded.

Our academic year had two semesters of 4 months (16 weeks) each. Physical chemistry was presented in two parts—Physical Chemistry I and II, each in one semester, four 1-hour lectures per week. This involved about 50 hours of lecturing and experimenting in each semester, a very comfortable time schedule for such a course. I had occasionally attended the lectures of W. Nernst in Berlin and of G. Bredig in Karlsruhe on physical chemistry and had a good idea of what experiments and demonstrations are particularly impressive and how to present the complicated laws of thermodynamics in a lucid and somewhat attractive manner. For the modernization of this "main course," two of my regular assistants, F. Patat and E. Suess, helped. If they had to work longer hours or if expensive instruments were needed, I always had enough funds to compensate them from my ongoing contract with I. G. Farben or from lecture honoraria that I received from seminars in industry.

A former colleague recalls his impressions of Mark as a professor:

> Mark usually had a broad smile, strictly parted hair (with strands of white even then), and a fashionable suit. Walking briskly into the lecture room as his weight shifted from one foot to the other, he would pause, smile, and begin lecturing. His style was slow, careful, and pleasing. If visitors were present who spoke a language other than German, he would pause periodically and outline his concepts in

> their language. He spoke English, French, and Italian in addition to his native German. His lectures were clear, cleverly constructed, enthusiastically delivered, and laced with samples and experiments.[1]

Interestingly, Mark was viewed as undignified.

> Each morning he would run up the steps of the main staircase to his office, pausing only to check the bulletin board on the half-way landing. Undaunted by this criticism, Mark founded and fitted out a football team for the First Chemical Institute. Then, he kicked with them. Mark, a former player for the Vienna Sports Club, played center-half for his team when it met the Second Chemical Institute in a big athletic showdown. The game was held at the Elektraplatz on Engerth Street, and although few recall who won, those who attended remember that Professor Spaeth, the Director of the Second Institute, observed Mark's exertions indignantly.[1]

Designing a Polymer Chemistry Curriculum

As soon as my teaching obligation was taken care of, I could concentrate on the really new and exciting academic task, namely, the design of a curriculum on polymer chemistry. It had become a new branch of chemistry, like electrochemistry and photochemistry before it, and, obviously, needed a carefully worked out special curriculum to put it on the same level with inorganic, organic, physical, and analytical chemistry.

At that time (1932) about a dozen laboratories cultivated polymer chemistry, most of them in industry. The leaders were W. H. Carothers at Du Pont, J. L. Swallow at Imperial Chemical Industries (ICI), H. Lecher in Leverkusen, W. Reppe (my successor) in Ludwigshafen, and F. Mayo at U.S. Rubber. None of them was interested in teaching, but only in basic and applied research for industrial purposes. There were also a few university laboratories where special fundamental work was carried out, such as by W. T. Astbury in Leeds, H. W. Melville in Cambridge, and G. V. Schulz in Staudinger's laboratory in Freiburg.

Even at the universities, there was no organized teaching schedule in polymer chemistry. Of course, Staudinger, W. Kern, and G. V. Schulz had many graduate students who worked for their Ph.D. degrees and wrote important and excellent papers, but there were no undergraduate courses or seminars. This situation raised a question for us in Vienna: How should a systematic curriculum for the teaching of polymer chemistry be included into the standard programs of undergraduate and graduate education?

This young branch of chemistry, which had developed like electrochemistry and photochemistry before it, demanded careful and serious consideration as a new scientific discipline. After lengthy discussions, my associates and I arrived at the conclusion that, for the beginning, the following courses should be offered in this new field:

- One two-semester course on Introduction to Polymer Chemistry
- One two-semester course on Organic Chemistry and Synthesis of Polymers
- One two-semester course on Polymers in Solution
- One one-semester course on Polymers in the Solid State

All kinds of changes were effected during a 4-year period of "trial-and-error," but, in general, the schedule was adequate for the time being.

Any good teaching requires textbooks or, at least, comprehensive review articles. Therefore, *Physical Chemistry of High Polymers* and *Polymers in Solution* were written or initiated rather soon. In this manner, my teaching responsibilities were gradually taken care of in an orderly fashion. The next really important question was what should happen to research.

Research Activities

Solution Viscosity. There seemed to be several areas where additional basic information was still badly needed; one was the significance of solution viscosity and its application to polymer systems. The question had its roots in a simple equation that Einstein had derived almost 30 years earlier. He found that the viscosity of a liquid (η_0) was increased by the addition of spherical particles of the volume fraction (v/V) to a viscosity (η) given by

$$\eta = \eta_0 + 2.5\eta_0(v/V)$$

where v is the aggregate volume of the spheres and V is the total volume of the system. This relationship applies if the particles are all the same size and are large compared with the molecules of the liquid, if $v/V << 1$, and if the suspended particles do not attract or repel each other.

Einstein applied his equation to solutions of sugar in water, assuming correctly that the dissolved and hydrated sugar molecules ($C_{12}H_{22}O_{11}$) plus five or six H_2O molecules are "large" in comparison with an H_2O molecule. We wanted an independent test with beads of microscopically visible size (10–100 μm) and essentially equal volume. Using glass beads suspended in aqueous salt solutions of the same specific gravity, H. Margaretha, G. Bunzl, and O. Goldschmidt carried out a doctoral thesis project, under F. R. Eirich's supervision, in which they varied the individual particle diameter, shear velocity, and total volume fraction and arrived at a linear plot for dilute suspensions

$$\frac{\eta - \eta_0}{\eta_0} \equiv \eta_{sp} = 2.4(v/V)$$

where η_{sp} is called specific viscosity, that was in excellent agreement with Einstein's prediction. As soon as v/V was larger than 0.05, the plot showed an upward curvature:

$$\eta_{sp} = 2.4(v/V) + 14(v/V)^2$$

A correction of this magnitude was later derived by R. Simha.

Having thus developed a reasonably practical model for the viscosity of dilute suspensions, we performed additional tests. These tests were carried out by several graduate students under the supervision of F. R. Eirich, R. Simha, and E. Guth, with rod and platelike particles for which theoretical expressions had been worked out by E. Jeffries and other investigators.

Condensation Polymerization. There were other areas of polymer science where additional experimental and theoretical work was needed to arrive at a complete and clear understanding of the various processes. One was the mechanism of polymerization, of which, at that time, three types were known: addition polymerization without a termination, such as the conversion of ethylene oxide into poly(ethylene oxide); addition polymerization with a distinct termination step; and condensation polymerization.

This third method had been thoroughly studied in the Du Pont laboratories by W. H. Carothers and his group for about 4 or 5 years. They had obtained some very interesting products—polyesters and polyamides.

In the wake of these classical studies on polycondensation reactions, P. J. Flory had established the rate equations for this important road to the synthesis of polyesters and nylons. At the same time, he gave an elegant formula for the molecular-weight distribution of these materials.

In our laboratory in Vienna, R. Raff and E. Marecek carried out some measurements on the polycondensation of a few aliphatic dicarboxylic acids and verified the Flory distribution by fractionation of the resulting polymers and end-group titration. Interesting work on the polymerization of vinyl compounds was done in various laboratories, particularly by G. V. Schulz in Freiburg. We felt that some experiments should be carried out in support of these theories. These experiments, performed by J. W. Breitenbach, confirmed the existence of distinct initiation, propagation, and chain-termination steps in such polymerizations.

Entropy and Chain Contraction. The behavior of flexible long-chain molecules had become a topic of intense interest in Ludwigshafen. Meyer, von Susich, and Valko had shown qualitatively that the driving force for rubber contraction was a tendency to increase the entropy. Now it was time to develop the necessary mathematical treatment. My own familiarity with statistics and its consequences was totally inadequate for such a task but, fortunately, Hans Thirring, professor of theoretical physics at the university, recommended Eugene Guth, one of his associates, to work on a solution for this problem.

Using a general equation that Lord Rayleigh had developed for the "random walk" of a particle, Guth deduced the number of conformations that would give a certain end-to-end distance smaller than the fully extended length of the chain. For a chain of n units of length b, the root-mean-square end-to-end distance is $bn^{1/2}$. For any given end-to-end distance, a certain probability W(random) exists, which is expressed as a function of the number n of units and the length b of each unit.

Some 50 years earlier, at the University of Vienna, Ludwig Boltzmann had related probability to entropy. The individual chain molecules of a stretched rubber band possess fewer conformations than the same chains in the contracted band and will, therefore, have a lower entropy and a tendency to contract. My work with Guth dealt with the elasticity of an isolated chain.[34] Later studies by many researchers expanded this work to interpret the behavior of cross-linked rubber networks.

Coiling of Chain Molecules. The coiling tendency of flexible polymeric chain molecules found another, quite different, application in the interpretation of their solution viscosity. In our laboratory, F. Eirich and R. Simha worked on that problem. According to Staudinger, the specific viscosity, $(\eta - \eta_0)/\eta_0$, of such systems is directly proportional to the product of the polymer concentration, c, and the molecular weight of the polymer, M. During our work with cellulose acetate in Ludwigshafen, we had found that this relationship was reasonably well fulfilled. However, as soon as we tried to apply it to rubber and polystyrene solutions, we found that the influence of the molecular weight as measured by osmotic pressure or diffusion was definitely less than proportional to η_{sp}/c.

Cellulose and its derivatives consist of relatively rigid molecules, whereas the chains of rubber and polystyrene should be much more flexible. In solution, they will have a tendency to coil. Because the end-to-end distance of a random chain is proportional to the 1/2 power of its length, the spherical volume occupied by the chain is proportional to the 3/2 power of this length. Thus, if the coiled chain behaves like a sphere in Einstein's theory, with a radius proportional to the root–mean square end-to-end distance, the effective hydrodynamic volume of these chains at any volume fraction increases as the square root of the molecular weight. Our measurements indicated an exponent between 0.5 and 1.0, and it was obvious that we should consider this exponent as a parameter that takes care of the intrinsic flexibility of the chains and of their interaction with the solvent.

At the same time, R. Houwink in Delft had carried out extensive tests with polyacrylate solution. He had formulated, independently, the relationship $[\eta] = KM^a$, where $[\eta]$ is η_{sp}/c extropolated to $c = 0$, with K and a constants characteristic of a polymer–solvent pair. This semiempirical and unsophisticated relationship was, for more than four decades, a useful tool for arriving at the approximate viscosity-average molecular weights of almost any kind of linear polymer. Today gel permeation chromatography allows us to determine not only the average molecular weight, but also the molecular-weight distribution.

Publicity and Contacts

Besides organized teaching and special areas of intriguing research, another job fell almost completely on my shoulders, at least in the beginning. I was responsible for making our laboratory known nationally and internationally as a new and upcoming factor in modern polymer science. At the time this discipline was almost unknown in Austria. It was of utmost importance for our reputation and future development to show that a new center of polymer science was in the making. This center would produce not only interesting new results but also, eventually, a new generation of polymer scientists who would be available for employment in the industry.

The well-known and, in general, favorably accepted results obtained in Dahlem and Ludwigshafen made a good background for such a propaganda drive. Publishing original papers and having seminars or symposia in Vienna were not enough. Traveling abroad and lecturing at international conferences had to become a routine activity. Such efforts started in 1932 when N. Semenov, who had been our guest in Dahlem for a few weeks, invited me to attend the celebration of

I made numerous journeys abroad, visiting laboratories and attending conferences. In 1953, the academician N. Semenov (right) and I again met in Russia, 20 years after my first visit in honor of Mendeleev's 100th birthday.

Second from left, I am greeting two distinguished guests who had come to tour the facilities at the University of Vienna, 1935. They are Professors Z. A. Rogowin of the Soviet Union and G. Saito of Japan. Also present were E. Guth, far right, and Lisl Mark Czitary, my sister, who still resides in Vienna.

Mendeleev's 100th birthday in Leningrad and Moscow. I stayed for 2 weeks in the Soviet Union. As a result, during the next 3 years several Russian scientists, Z. A. Rogowin, N. Neijnan, K. Frumkin, and A. V. Kargin, spent several weeks in our laboratory. A friendly contact with Soviet science was established.

This contact had an interesting sequel. After the discovery of heavy hydrogen by H. Urey in the late 1920s, Professor Eucken invited me to Göttingen to give a lecture on this topic. During the discussion it was mentioned that because HDO and D_2O have higher melting points than H_2O, heavy water should be concentrated in old glaciers. I first went with Eirich to the Jungfraujoch to collect samples from the glacier, but we could find no difference between the density of these samples and that of ordinary water. I suspected that the Jungfrau glacier is too young to show the effect with our experimen-

During the years 1934 through 1936, the institute was charged with the responsibility of investigating the heavy water content of glaciers. The first trip to make such an investigation was to Switzerland. On the far right is Professor von Laue, next to him is Professor Ladenburg. I am on the far left.

tal precision. When I consulted my geology colleagues, I was told that the Bezinghi glacier on the Dych Tau in the Caucasus is very old. I wrote to Professor B. Vavilov of the Soviet Academy of Sciences and got an invitation to participate in a joint expedition in August 1935. I was accompanied by E. Baroni, who later reported our results. As it turned out, analyses of samples collected on the Bezinghi glacier failed to indicate an enrichment with heavy water. However, on the top of the Elbrus (at an altitude of 5630 meters, the highest in Europe) a sample collected 20 cm below the surface had a deuterium content 25% higher than that in freshly fallen snow. Of course, I also used this wonderful opportunity for some additional mountaineering, climbing the Ushba.

After the first year, I also started to reestablish personal contact with universities and industrial organizations in the West. Since the Ludwigshafen days, we had frequent and close relations with groups in England (Universities of London, Manchester, Cambridge, and Birmingham, and ICI, Courtaulds, and

In 1935 the search for the concentration of heavy water in glacier ice was extended to the Caucasus mountains. Here is part of the caravan carrying equipment and food into the high mountain region.

On the summit of the Ushba mountain, one of the most picturesque peaks of the entire Caucasus range, 1935.

British Rubber Producers in Welwyn Garden City); in France in Paris, Nancy, Strasbourg, and Clermont Ferrand; in Belgium in Louvain and Liège; and in Holland in Leyden and Utrecht. Whenever possible I took along one or two of my advanced assistants—F. Patat, F. Eirich, E. Suess, E. Valko, R. Simha, E. Guth, E. Broda, and R. Raff—to present them as upcoming young representatives of polymer science and to initiate their future international relations with the leaders of chemistry and physics in the West.

In a very few years it would become clear how important the personal contacts were for them and, for that matter, for myself. During these visits, I usually gave a few lectures. Our institute slowly became known as a new and significant center of polymer science. Since my departure from Ludwigshafen, I visited Germany regularly as a consultant of I. G. Farben and as a participant of scientific societies such as the Bunsen or Colloid Society.

Concept of Macromolecules

Naturally, if you look back on a somewhat long period of the past, it simplifies matters if you single out certain events of special importance. One such event was a farewell lecture given by Staudinger in 1926, when he left Zürich to become a professor at the University of Freiburg in Germany. With great skill and vehemence Staudinger presented his ideas about the existence, structure, and properties of macromolecules, but he met with little support and much forceful opposition.

Resistance to the Concept of Macromolecules

There were essentially three groups of scientists who did not like Staudinger's new concepts and conclusions. The first objections came from his own closest friends and colleagues, classical organic chemists like Heinrich Wieland, Adolf Windaus, and Paul Karrer—all Nobel Prize winners. They would summarize their position in the following way: "We have worked all our lives with organic molecules in the molecular weight range 200–800, had many difficulties in identifying their structure and, eventually, accomplished their synthesis. The statement that there exist organic molecules having molecular weights of several hundred thousand, some soluble like rubber or starch and others insoluble like cellulose or silk, creates a 'credibility gap' between your methods of preparation, purification, and identification and our own." One of them commented, jokingly: "Suppose you are a zoologist and make an expedition to Central Africa. After your return you give a lecture and say, 'In the plains and steppes of the Congo, I have seen elephants that are 80 meters long and 20 meters high.' Nobody is going to believe you, even if you have a few long-range snapshots. You must have actual 'end-to-end' measurements with an expandable measuring tape." This Staudinger did not have for his macromolecules. Altogether the criticism of this group was essentially subjective, but in view of the high distinction of its members—half a dozen Nobel laureates—they carried much weight.

Another group of equally prominent scientists (J. Perrin, T. Svedberg, W. Ostwald, E. K. Rideal, and H. Freundlich) responded by saying, "We know your systems well; they are colloids. We do not question the existence of many objects of this size, but we cannot accept your insistence that they all are held together by covalent bonds. We have good evidence that these aggregations, agglomerations, or clusters of small organic molecules may be held together rather firmly by various associative forces—polar groups, complex formations, and others. For us, the proclamation of a macromolecular science is not necessary. We consider it as a part of colloid science, and we know how to classify and explain its phenomena." For this group, the Staudinger theory was not *unbelievable*; it was *unnecessary*.

The position of the third group—the crystallographers—was dramatically expressed by Paul Niggli, one of their preeminent mentors. He said: "It has been established that the crystallographic elementary cell of cellulose, silk, rubber, and related natural materials is small, having space only for three to five chemical units. Because the molecule cannot be larger than the basic crystallographic cell, the building units of all these compounds must be small." Thus the criticism went from *incredible* to *unnecessary* and to *impossible*, a nice spectrum of opposition to a new theory. Staudinger remained firm and essentially supported his ideas and concepts by solution viscosity measurements, which, however, made not much of an impression on his audience.

In the mid-1920s several meetings of this character took place. They were all interesting and stimulating, but inconclusive. Frequently they were not only controversial but also aggressively critical.

Even the champions of the long chain aspect did not agree with each other, as they easily could have done; because instead of concentrating on the essential principle, they disagreed on specific details and, on certain occasions, they argued with each other more vigorously than with the defenders of the association theory.[1]

In the summer of 1926, Professor Haber called me into his office in Berlin-Dahlem. After friendly greetings, he told me the following: "You have occupied yourself in the Fiber Research Institute for about 4 years now, making the complete quantitative structure examinations of normal organic substances, and have made quantitative investigations with cellulose, silk, and rubber by using X-ray techniques. The Society of German Naturalists and Physicians (the German equivalent of the American Association for the Advancement of Science) has its annual meeting in Düsseldorf in September 1926. The Chemical Society intends at this opportunity to organize a general discussion about the constitution of the so-called high-molecular-weight organic substances. We would like you to give a lecture on the general topic of X-ray structure examinations of organic substances and especially to explain whether a small crystallographic elementary cell excludes the presence of a very large molecule."

Haber was, as usual, right to the point and had understood the situation very clearly. Most of the arguments pro and con were qualitative, even subjective. On the other hand, the small volume of the elementary cells of cellulose and silk had been well established; each of them could only contain four units. Now, if it were true that a molecule could not be larger than the elementary cell established by crystallographic investigation, there would be a clear-cut and objective argument against the existence of very large molecules.

On September 23, 1926, Richard Willstätter opened the famous meeting in Düsseldorf and called the first speaker, Max Bergmann, who presented several arguments for the aggregation structure of insulin and certain proteins. He used these arguments, together with a small elementary cell determination for silk, to refute Staudinger's macromolecular postulate. Hans Pringsheim proceeded in a similar vein, referring to insulin and a few other polysaccharides. Both scientists quoted P. Karrer, K. Hess, and R. Pummerer and referred to the well-known Werner complex compounds and to the high viscosity of many colloidal solutions. Staudinger presented extensive material on rubber and some synthetic polymers and based his contentions essentially on the high viscosity of polymer solutions. He also disproved the theory advocated by Carl Harries, which had been widely accepted, that rubber consists of small molecules held together by "residual valence forces" between C=C double bonds, by showing that the "colloidal properties" of rubber remain unchanged when the double bonds are eliminated by hydrogenation. He had no data on cellulose or proteins but suggested that they also have a macromolecular structure.

Gradual Acceptance

The principal objection to Staudinger's proposal was the small elementary cell, and that was just the topic about which I had to speak. I pointed out that there are cases, clearly discussed in several articles by A. Reis and K. Weissenberg, where the molecule can be larger than the elementary cell. This is always the case when true chemical main valences penetrate through the whole crystal. This happens in two directions in graphite and in all three directions in the diamond structure. Generally, this means that small elementary cells obtained by X-ray analyses do not exclude the presence of large molecules, but they also do not prove their existence. It is a pity that neither the discussions nor the concluding remarks of Willstätter have been published. As far as I can remember, Willstätter thanked all lecturers and discussion speakers in friendly words and said: "For me, as an organic chemist, the concept that a molecule can have a molecular weight of 100,000 is somewhat terrifying, but on the basis of what we have heard today, and of additional quantitative data that may be forthcoming, it seems that I shall have to slowly adjust to this thought."

Unquestionably more than any other single observation or measurement, the brilliant studies of W. H. Carothers and his associates (initiated in 1928 in the Du Pont Experimental Station in Wilmington, Delaware) contributed to the ultimate breakthrough in favor of the long-chain (macromolecular) concept. While working with polycondensation products of polyester and polyamides, Carothers demonstrated that chain molecules consisting of many hundred monomers may be synthesized. Their "degree of polymerization" (DP), the number of chemically combined units, was quantitatively determined by the titration of end groups. The essential requirement for the synthesis of high-molecular-weight condensation products is extreme purity of the two components (diacid and diamine or glycol) and great care in the removal of water during the polycondensation reaction.

Together with the execution of the crucial experiments, the theory of these processes was also developed in Germany by G. V. Schulz and in the United States by P. J. Flory. All of this research led to a new level of information and understanding that influenced the spirit and atmosphere of another historic conference in Cambridge, England, on September 26–28, 1935, under the sponsorship of the Faraday Society and under the chairmanship of E. K. Rideal. This time the existence of macromolecules was accepted as an established fact and the discussions revolved about their synthesis and properties.

The high point of the meeting was the presence of Carothers and his report on systematic and fundamental work on addition and condensation polymerization, which he had started with a group of about 10 associates in 1928 at the Du Pont Company. Everybody who was actively engaged in polymer research was there: H. K. Meyer and G. V. Schulz from Germany, T. Svedberg from Sweden, R. Katz from Holland, G. Champetier from France, and representatives from at least five other countries.

At that time there was no more doubt about the existence of macromolecules, but considerable difference of opinion about their structure and molecular properties. In order to explain the high specific viscosities of macromolecules, Staudinger postulated that, even in solution, those molecules behave like "rigid rods". This idea was evidently inconsistent with the well-established principle of hindered rotation about a single covalent bond, as in ethane and polyethylene. To demonstrate his concept to an audience, Staudinger always had with him bundles of match sticks, usually about 10–20 inches long. (Staudinger's match sticks can be seen in the Deutsches Museum in Munich.) He would say, "Those match sticks represent macromolecules in solution and are responsible for the high viscosities of these systems." I recall that, at the Faraday Society meeting in Cambridge in 1935, he showed such sticks, broke one, and said, "If molecules are degraded, their strength is reduced and the viscosity drops dramatically." Then, turning to E. K. Rideal, who was in the chair, he said, "Professor Rideal, what I just said—isn't it clear?" Rideal looked at the broken sticks and said calmly, "Yes, I think it is perfectly clear, but wrong." That terminated any further discussion of the point.

Nazi Threat and Subsequent Emigration

Having lived in Germany for 10 years, I belonged to several social circles and had many friends, colleagues, and even relatives. It was obvious to me that things were moving from bad to worse. Conditions developed toward complete conformity exactly as the Fuehrer had foreseen in his book *Mein Kampf*. Most disturbingly, where there was not immediate and complete subservience, increasing force and brutality were used in all domains: art, science, religion, education, and even sports and relaxation.

Back home in Austria, conditions were also disintegrating from day to day. Mussolini had reacted to his diplomatic isolation during his conquest of Ethiopia by tying Italy ever more closely to Germany, so that he was no longer able to prevent the Fuehrer's annexation of Austria as he had been able to do in 1934. The "Austro-Nazis" became increasingly aggressive, blowing up public telephone booths, burning cars that were owned by Jews, and preventing Jewish professors from giving their courses at the university. Nothing could be done against these terrorist acts.

At the institute, I used the 1937 vacation to complete work on quantitative fractionation with G. Saito, to construct an improved osmometer with H. Höhn, and to study cellulose degradation with Z. A. Rogowin. One day in midsummer 1937, I received a letter from C. B. Thorne, technical director of the Canadian International Pulp and Paper Company, one of the largest wood-pulp-producing companies in the world—producing about 1500 tons per day. In spite of this volume, they had no real basic understanding of the various production steps; everything was done by trial and error. Thorne visited Europe every summer and wrote that he would like to meet and discuss their problems with me. I brought the letter with me to the Canadian Embassy in Vienna to find out the exact address for my answer. The officer at the embassy was extremely polite. Of course they knew the company and even Dr. Thorne, who was Norwegian by birth and visited Europe every summer. When I said that I would like to thank him and send a cable about time and place to meet, the officer volunteered to do it for me because "in Austria, cable service abroad may already be restricted or, at least, controlled."

In any event, after several cable exchanges, Dr. Thorne and I met in his hotel in Dresden in September 1937. He told me: "We have a large production of wood pulp for making rayon, cellulose acetate, and cellophane without adequate support of basic knowledge of our process. We have a research laboratory, but it is antiquated and obsolete. We need modernization of methods, equipment, and personnel. You have worked successfully in this field in a large research institute in Berlin and later in an industrial development laboratory at I. G. Farben. I am offering you the position of research manager at our laboratory in Hawkesbury, Ontario."

I thanked Dr. Thorne profoundly for his generous and most attractive offer and said, "Right now my teaching and research activities at the university are taking all my time, but I could try to come to Canada next spring and summer on a temporary basis and work out a plan for a reorganization of your laboratory." He was not quite satisfied but agreed that this would be a possible way to proceed. I also gave him the names of two other candidates, H. Kraessig and G. von Susich. After his return to Canada, I sent Dr. Thorne a list of modern methods and instruments that would be particularly desirable. He thanked me and put me in touch with his deputy director, Sigmond Wang.

The next few months were turbulent and trying. My main course was now to help my associates to get out of Austria because they all were in the danger zone. In a few cases, it worked rapidly and easily—P. Gross, M. Perutz, and H. Motz went to England; W. Hohenstein and E. Suess to Italy; others to France and even to the United States. When I suggested to Perutz that he join the famous X-ray laboratory at Cambridge, he objected that he knew nothing about this discipline. I told him: "You will learn." And he surely did. He received the Nobel Prize in 1962 for establishing the crystal structure of hemoglobin.

Once I had addressed the safety of my friends, I began preparations for my own leave-taking. Early in 1938 I started delegating my administrative duties to colleagues who had decided to remain at the institute. One associate recalled, "Mark left a well appointed and orderly house".[1]

The word "Hawkesbury" never left my mind and, in fact, in the end it provided an escape route for me. Everything—phone calls, telegrams, and letters—was somehow controlled. I knew that the superintendent of our house and several students belonged to the Nazi storm troops. They observed and reported everything we did or said. My wife and I knew we had to leave Austria, and we knew that we would not be allowed to take any money or valuables with us. We needed 4 months of careful and clandestine action to buy platinum wire, which we bent into coat hangers. Mimi knitted covers for these coat hangers so that we could take them out safely.

When Hitler invaded Austria on March 11, 1938, we were at home in Vienna. The next day I was arrested and put into the Gestapo prison in Vienna, where I was interrogated for several days, mainly about my relations with Dollfus and I. G. Farben. Stripped of my passport, I was released with a stern warning not to have contact with anyone Jewish.[1] I went straight to the Canadian embassy and sent a cable to Hawkesbury that I was now ready to come immediately. With the aid of some bribing (an amount equal to a year's salary), I got my passport back and, at the embassy, the visa to enter Canada. With this visa in my passport, I had no difficulty getting transit visas through Switzerland, France, and England.

At the end of April we (Mimi and I, the two boys, and my Jewish niece, Greta Kraus, a prominent harpsichordist) mounted a Nazi flag on the radiator of the car, strapped ski equipment to the roof, and drove from Vienna, safely arriving the next day in Zürich, where I visited our friends, Professors Leopold Ruzicka, Paul Karrer, W. Kuhn, and P. Scherrer, to ask them for help and assistance for our colleagues who were still in Vienna. We went to England via France—to London, Birmingham, and Manchester, where I did some work with the Shirley Institute on synthetic fibers. My family used this time to improve their English and to visit friends, mostly emigrants from Germany. In mid-September my family drove to Liverpool to see me off when I boarded the boat to Montreal, then returned to London for a few more weeks to await the date of their departure. On September 26 I arrived in Montreal and a few hours later stopped my car in front of the main gate of the Hawkesbury Mill.

During the crossing to Montreal from Liverpool on the *Duchess of Richmond*, a slow boat, I had ample time to almost finish the English edition of my *Physical Chemistry of High Polymers.* I had also received an invitation from Professor H. Hibbert at McGill University to present some lectures on the structure of cellulose. The passage to Canada gave me an opportunity to draft these talks. I suspected that I would accept an academic position once the Hawkesbury operation had been modernized, and Thorne understood that this was my intention. Also, the sea voyage provided the right atmosphere for me to reflect upon some of the high points that I had witnessed in the development of polymer science.

> Mark's open feelings about those who drove him out of Austria in 1938 are curiously free of contempt. He describes the Nazis as "misguided" and those scientists who supported them as "unfortunate". . . . After the Second World War, Mark was, in fact, very active in the reestablishment [of] German and Austrian scientists to the world scientific community. His first action on returning to Vienna in 1947 was to call on his indirect successor at the First Chemical Institute, Professor L. Ebert, and reassure him that he would "never attempt to drive him out of the position he filled in such an excellent manner".
>
> . . . One coworker of more than thirty years recently commented regarding Mark and his family's flight from Austria in 1938, "I knew he left, but I never realized the circumstances. I guess I always assumed they (the circumstances) were the best possible since the *Geheimrat* has never complained. He doesn't seem bitter, yet he has a right to be."[1]

Canada

1938–1940

A few days after my arrival, Mr. Sigmond Wang, the manager of research at the Canadian International Pulp and Paper Company, introduced me to his research team, a number of relatively young people very well trained in classical chemistry (organic, physical, colloid, and analytical). None of them knew much about polymer science—molecular weight and MW distribution, orientation, crystallinity, fiber structure, and their influences on the technical properties of the resulting fibers, such as tensile modulus and strength, elongation to break, loop and knot strength, and moisture resistance. My assignment was to span this bridge. The group of associates was very international; its members came from Canada, the United States, England, France, Germany, and Norway. Everybody spoke English with some kind of accent, but there was never a serious difficulty in communication.

Pulp Production

During the first conference, I was familiarized with the problems that should be solved in the near future. In two mills, Hawkesbury and Temiskaming, the Canadian International Pulp and

Paper Company made both pulp for the paper industry and two types of "chemical pulp"—one for the production of cellulose acetate and one for the manufacture of viscose rayon–cellulose xanthate. Our main customers for the cellulose acetate pulp were Du Pont, Celanese, and Eastman Kodak; those for the viscose rayon type were Du Pont, Beaunit, and Aviso.

The acetate pulp was used to manufacture photographic film and acetate filaments, both requiring an extremely pure, uniform, and reactive raw material. The bulk (about 80%) of our wood pulp went into viscose rayon fabrication, staple fiber, and continuous filaments. During the few years preceding 1938, our customers were generally satisfied with our products, although two pulp-producing competitors (the Rayonier Company on the West Coast and Brown Pulp and Paper Company in Maine) kept us permanently on our toes. In 1937, Du Pont had introduced a new use for viscose rayon filaments in the manufacture of tire cords. In the past such cords had been made from a special type of cotton, but the Du Pont people felt that, with the present advanced knowledge of cellulose structure and viscose rayon spinning, a better cord could be made from adequately "tailored" wood pulp.

I knew the cotton structure and texture well from our work in Dahlem; its only advantage over wood pulp was that its molecular weight was much higher than that of any presently used wood pulp. The degree of polymerization of cotton was at least 1500, whereas that of commercial wood pulp was between 300 and 1000. This gave cotton an important advantage with respect to tensile strength. On the other hand, the texture and structure of cotton was not favorable for cord properties, the degree of orientation was low because of the helicoidal structure of the primary wall, and the degree of crystallinity was on the low side (50–70%). Thus, from the point of view of DP, cotton was good; from the point of view of structure and texture, it was inferior. Toughness and durability of cotton cords were good, but modulus and tensile strength were unsatisfactory.

Evidently a new compromise between DP, structure, and texture was desirable. Our first step was to prepare a xanthate solution of cotton and spin it under tension in order to arrive at high orientation and crystallinity. Several difficulties had to be overcome. The viscosity of a cotton linters solution (DP around 1500, cellulose concentration normally around 10%) was too high for the solution to be filterable and spinnable at reasonable rates. The next step was, therefore, to reduce the DP by careful acid degradation, to prepare spinning solutions of reduced concentration so that, eventually, they could be processed under commercial conditions, and to spin them under high tension.

After a series of tests under statistically controlled conditions (today one would use a computer), we found that linters with a DP of 1200 and a xanthate concentration of 8% could be spun conveniently and gave a tensile modulus of 140 g/den, a tensile strength of 7–8 g/den, and a 15% elongation to break. This result was distinctly better than whatever had been obtained with cotton cord or with any rayon used before.

Our next problem therefore was to design a new type of wood pulp (cordicell) that would match these figures obtained with artificially adjusted cotton linters. It was necessary to juggle a number of parameters (namely DP and DP distribution of the pulp, concentration of the spinning solution, conditions of spinning and drying, comparative speed, tension, and ultimate relaxation). Because we were able to measure each parameter quantitatively, the solution of this problem was possible, but only by slow and careful systematic work. This work was started immediately. After a few weeks we had laboratory samples of a pulp with quite promising properties: a modulus of 120 g/den, a tensile strength of 8 g/den, and a 12% elongation to break.

A few additional adjustments were made, mainly connected with the yield of the pulp based on "bone dry" wood. After a few weeks we went into our pilot plant and began making half a ton of pulp per day. The start was somewhat shaky, but we got a pulp that looked good for tire cord production. We sent an adequate quantity to Du Pont. Two weeks later, we got their reaction: "We like this pulp very much. Now we need larger quantities." That is the way it always goes. Until Wednesday evening, nobody knows about or wants your product. By Thursday morning, everyone urgently needs tons of it.

As a result, a few weeks later (with trepidation) we made a plant run in the Hawkesbury Mill. Luckily, we came up with an acceptable pulp—not as good as we had hoped for and not as bad as we had feared.

Matters were now in the hands of our engineers who, for

many years, had been accustomed to converting products and methods from the research domain to a large-scale application. They succeeded in streamlining the process for the "Kippawa" Mill, Temiskaming, making 300 metric tons of pulp per day. This huge quantity would fill about 20 boxcars containing 15 tons each. This "dissolving" or "chemical" pulp was sold at a much higher price than the various types of "normal" paper pulp. Its price was determined by about a dozen specifications such as water content; α-, β-, and γ-cellulose; lignin; resin; ash; and solution viscosity.

With actual production we faced another problem: How many test samples of what size must be analyzed every day to arrive at a credible characterization of the daily 300-ton shipment? We analyzed about 100 samples per day, which were taken according to a special schedule, each sample weighing about 100 g. Less than 1 ppm of our merchandise was tested for moisture content, for example. Was this really sufficient to characterize the whole lot? The analytical people belonged to my research group, but this kind of testing was routine, so I did not touch it and left it as it was.

A more interesting and appealing problem was the modernization of our "acetate pulp." What was essential here was to have a very low content of impurities such as lignin, resin, ash, and dust, together with an almost complete absence of γ-cellulose and a high reactivity for the acetylation process. We modified the standard tire cord pulp by an additional alkaline extraction of γ-cellulose and an additional treatment with chlorine to increase whiteness and reactivity.

Monograph Series

Thus the days in Hawkesbury were filled with interesting projects, all of which amounted to the careful application of recent fundamental knowledge to practical production procedures. Much of my time was also devoted to initiating the publication of a journal on polymer chemistry in America, as we had done in Vienna a few years earlier, because the United States also lacked systematic polymer teaching and research at academic institutions. Industry, on the other hand, was very active and successful, including work at Du Pont, Monsanto, Dow, and Allied, among others. G. S. Whitby and I thought, therefore, that a volume on W. H. Carothers' work at Du Pont (the "hero" of nylon and neoprene) would be a good beginning for a monograph series on polymers. I started working on it in 1938, together with M. Dekker and E. Proskauer of the Interscience Publishing Company. The volume appeared in 1940. Now the monograph series "High Polymers and Related Substances" contains more than 40 volumes.

Return to Academia

The modern instruments and advanced equipment had arrived in 1939, and the individual members of the research department were well familiarized with their application. Hence, what Dr. Thorne had asked me to do was essentially achieved, and I wanted to return to an academic institution. By mid-1940 Canada was already at war with Germany and I was technically and officially an enemy alien. Therefore, a position in the United States would be preferable. It happened that one of our Du Pont contact scientists, W. F. Zimmerli, was a board member of the Polytechnic Institute of Brooklyn. On his advice the president of "Poly" offered me a position as adjunct professor for fall 1940.

When I told Dr. Thorne of my intention to leave, he was not very happy. However, as soon as he understood that I would become a technical consultant of Du Pont, he realized that now, after the modernization of the Hawkesbury Laboratory, I could be more valuable to his company's chemical pulp business in the United States than in Canada. In the summer of 1940 (at the height of the war in France) we drove from Hawkesbury to New York City. This was a difficult transition for our two sons, who loved life in the Canadian countryside and were particularly excited about going to school on skis in the winter.

Brooklyn in the War Years and Early Post-War Period

1940–1947

Shellac Bureau

When I arrived at the Polytechnic Institute of Brooklyn in September 1940, professor R. E. Kirk, head of the chemistry department, assigned me to the "Shellac Bureau". This group had existed at the Polytechnic Institute for several years under the directorship of William H. Gardner. Its activities had been essentially the testing and control of incoming shipments of shellac from India and Indonesia, plus some advanced organic chemical research on the composition and structure of various types of shellac and related resins for coating and electrical insulation. It was sponsored and supported by the U.S. Shellac Import Organization. When I arrived it was obvious that the war in the East would adversely affect the safety of shipments to Europe and the United States. Gardner was actively looking for an adequate replacement (natural or synthetic) to keep the industry in the United States properly supplied. During my years in Ludwigshafen I was well informed about synthetic resins with properties similar to and even superior to those of shellac. During my consulting years with I. G. Farben (1932–1938) I had kept up with the progress in this field.

Several vinyl ester polymers (acetate, propionate, and butyrate) together with polyacrylic and polymethacrylic esters (methyl, butyl, cyclohexyl, and others) were close to natural shellac in many respects. Specifically, from 1928 to 1932 we had carried out work on the synthesis, characterization, and application of such systems in other units of the company, Farbwerke Hoechst and the Wacker Chemie AG. They had worked out useful copolymers and polyblends, not only for the replacement of shellac but also for the preparation and production of an entire family of soluble resins. These resins were appropriate for a wide range of applications important to the shellac industry.

It was a lucky coincidence that I was able to transfer from Germany to the United States a science and technology that was interesting and valuable for my new employer. The many able and extremely helpful colleagues at the institute greatly facilitated the integration of polymer science and engineering–my new and special field of experience–into the mainstream of scientific and practical efforts in Brooklyn.

First of all, I. Fankuchen was an internationally famous X-ray specialist who came from Cambridge, England. Also, P. Spoerri, an organic chemist; Donald F. Othmer, a chemical engineer; Ernst Weber, an electrical engineer; P. P. Ewald, a physicist; K. G. Stern, a biochemist; and C. Waring, a physical chemist, were very helpful and encouraging. Most important was the helpful attitude of Harry S. Rogers, president of the Polytechnic Institute, who conducted us with great skill and success through the difficult war years.

My own responsibilities were essentially to equip two laboratories at the Shellac Bureau for experimental research in the polymer field and to start organizing courses that later would serve as a basis for a regular curriculum of polymer science.

Thus all my responsibilities revolved around research and teaching. Three things were most important for any progress in these fields: space, equipment, and people. With the aid of Professors Gardner and Kirk, together with support from industry, we immediately occupied two small laboratories of the Shellac Bureau, purchased some equipment, and hired a few assistants. As topics for these initial research efforts we would have to select areas that had been somewhat neglected until now in the United States—fractionation of natural and synthetic polymeric systems and quantitative determination of the molecular weight of the fractions. A good deal of such work had already been done in Vienna and Canada through the combined application of viscometry and osmometry and through the systematic use of the intrinsic viscosity–molecular weight relationship (mentioned earlier). We had experimentally verified this relationship for several polymeric systems, polystyrene, polyacrylates, vinyl ester polymers, rubber, and synthetic elastomers.

In 1940, a complete clarification of polymolecularity and absolute molecular weight measurements was an important research project. Another pioneering activity was directed toward an improved understanding of polymerization mechanisms by condensation, ring opening, and direct addition using various types of catalysts. Interesting related results had been recently obtained in several laboratories (e.g., in Vienna by H. Dostal, R. Raff, and E. Suess, and also in Germany, England, and particularly in the Du Pont laboratories by W. H. Carothers). In December 1941, when the United States entered the war, additional problems were added, particularly the permeability and impact strength of thin films, fundamental studies of emulsion polymerization, and phase transitions in polymer blends.

All these wartime activities were sponsored, financed, and controlled by the Office of Scientific Research and Development (OSRD). Our goal was to obtain useful results as rapidly as possible. Later, after additional tests and theoretical considerations, our findings were all published in the appropriate journals and as chapters in books that appeared after the war. Throughout all this work we were greatly aided by our two supervising officers, Captain William Aiken and Lieutenant George James, both with the Army.

The OSRD project allowed us to hire P. M. Doty, B. H. Zimm, and A. V. Tobolsky, all of whom later became leading figures in polymer research. Tobolsky also helped us in the teaching of polymer courses. My student, T. Alfrey, and Tobolsky's student, R. B. Mesrobian, joined our polymer faculty after receiving their doctorates. Later the staff of our Polymer Research Institute was strengthened by the addition of F. R. Eirich, D. M. McLaren, H. Morawetz, G. Oster, and R. Ullman. R. S. Stein, today a leading polymer physicist, also had his first exposure to polymer research while he was an undergraduate at Poly.

War Projects

During the war our group became involved in a few projects that had little or nothing to do with polymers but were considered potentially very important for certain problems of the war effort. These projects originated from experience I had obtained during my activities as a professor at the University of Vienna (1931–1937) and at the Canadian International Paper Company in Hawkesbury (1938–1940).

Refining the Weasel. In spring 1942 I received a phone call from C. P. Putnam, project leader at the OSRD in Washington, DC. "Are you the Herman Mark," he asked, "who made measurements of the shear strength of snow and published some papers on the content of heavy water in the glaciers of the Alps and the Caucasus Mountains?"

> Note: Putnam was referring to that period during 1932 to 1935 when Mark had returned to Vienna and to the mountains he knew so well:
>
> Unable to resist their call, he [Mark] reimmersed himself in their activities. Within a few months he was skiing regularly, serving on the local avalanche rescue squad, and conducting a personal, yet thorough, study of the causes of avalanches. Throughout the winters of 1932 and 1935 he dashed to the scenes of avalanches to question inhabitants about the conditions—weather, noise, stacking—just before the snow started to slide.
>
> The Ministry of the Interior soon recognized his growing expertise. In the winters of 1933–35, he was asked to serve on a committee charged with recommending road sites. The committee was made up of a mountain guide, an Army officer, several Alpine troops, and Mark, who served as the scientific advisor. The results of their study was used by the Ministry and Army when constructing new roads. Independently, Mark published the findings as part of a community report in an Austrian journal, thus establishing himself as an expert in the inexact art of avalanche prediction.[1]

When I said yes, he suggested that I come down to Washington and meet him in the library of the National Academy of

Sciences. The next day he told me the following: There would probably be some military action in snow-covered mountains—Northern Norway, Sweden, and Finland, and later perhaps in the Alps. The American army was developing an armed snowmobile (code name "Weasel") that was essentially a light, easily maneuverable tank to be used in surprise attacks on poorly accessible objectives such as radio stations, artillery observers, and flak positions. The Studebaker Company won the project and had completed half a dozen prototypes. These prototypes had been tested on their proving grounds and were now to undergo additional crucial field tests in snow-covered mountains.

The Columbia Icefields in the Canadian Rockies were chosen for these tests, to begin within a few weeks. The troops selected had some experience in mountain climbing, but they were not aware of the dangers that existed on high (above 14,000 feet) snow-covered peaks, such as avalanches, crevices, and ice breaks. In addition, the engineers who designed the Weasel needed information on the mechanical properties of snow and ice in order to develop criteria of conditions under which the vehicle could be operated without danger of it being destroyed by an avalanche or by tumbling down a steep icy slope. Mr. Putnam, a man of quite unusual intelligence, energy, and character, made a few additional remarks. He then asked me whether I would be ready to join his forces. I said yes and asked only for permission to take an assistant with me. This request was granted, and I returned to Brooklyn to make my preparations for this new assignment.

It was immediately clear that I should take Alfrey with me as my assistant, even though he was not a mountain climber or snow expert. His physical strength was more than sufficient for the task, and he could absorb the small amount of literature that existed in the field in a few days. The Columbia Icefields are a vast, relatively even, snow- and ice-covered plateau surrounded by a half dozen higher peaks, some gentle, some steep. We arrived together with the military detail and the Studebaker engineering group. While they were setting up the base camp, Alfrey and I started to get familiar with our territory. Alfrey had never been on skis before, but after a few days he was able to move around freely and to handle all necessary slopes. For training we climbed some of the surrounding peaks and eventually obtained a good survey of where the Weasel would have to be tested for its various performances. A big shallow trough was eventually selected for the very important and critical endurance test.

For Alfrey and me, the important object was snow and its impact on the possible military actions of the Weasel. Snow consists of two components (air and water) and of three phases (gas, liquid, and solid). Alfrey devised a rapid method of determining its composition: fill a beer can of known volume with snow and weigh it with a spring balance. This gives one the air content. Then add a known amount of hot water of known temperature and measure the temperature of the contents of the can. This gives you the water-to-ice ratio. These two simple observations provide a characterization of the snow sample. We determined the shear strength of many different snow types, and Alfrey drew a nomogram relating the shear strength to the snow sample composition.

The next variable was the inclination of the slope the Weasel was supposed to climb. Working with a test vehicle, we measured the shear stress it produced at various inclinations. In the end, Alfrey generated a two-page manual that would tell the driver of a Weasel whether or not it could successfully climb a slope covered by a given type of snow. Test climbs on the peaks surrounding the ice fields confirmed the practicality of this procedure. This work proved valuable when the Weasel was later used in action in Northern Europe. Mr. Putnam was highly satisfied with Alfrey's manual. As we were about to return to Brooklyn, Putnam said, "Come with me to Boston; I have another job for you."

Developing the Ducq. The new problem involved an amphibious landing craft (code name "Ducq") that later (in many variations) became much more important for the war effort than the Weasel did. The following questions were posed to us: What wind and waves could the craft handle in open waters? Through what kinds of surf could the vehicles land safely and rapidly? The literature on wave formation, propagation, and interference is enormous, but, fortunately, there existed, even then, a few handbooks that summarized the essential facts. Alfrey and I went to the oceanography section of the Weidner Library at Harvard University and surveyed the literature. It was not too difficult to arrive at criteria for what conditions (wind, swell, and waves) the various types of Ducq could safely be released in from transport ships in open water (i.e., about half a mile off the coast). Alfrey, together with two Navy officers, helped in formulating a few rules for this phase of the operation.

A series of tests were made under different weather conditions around Cape Cod. The fixing of conditions under the surf and for moving onto the beach and further inland was much more difficult and needed considerably more testing.

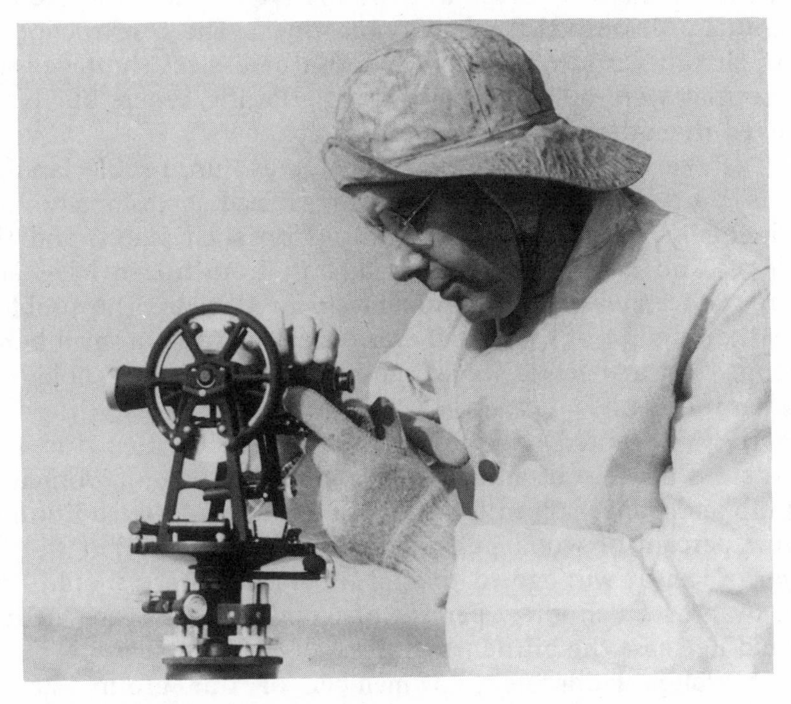

Using a Theodolite to take measurements on a landing craft in 1942 during my research to determine if icebergs could be used as landing fields.

Alfrey went with a large Navy contingent to Virginia Beach, where these tests were initiated and continued for several weeks. I returned to Brooklyn and to my responsibilities at the institute. Later, in the fall, Alfrey and I went for a few days to Cape Cod for some more advanced tests of the Ducq before several models went into mass production.

Building a Better Iceberg. In mid-1942 I received a phone call from John D. Bernal in Cambridge, England, asking me to get in touch with his colleague Geoffrey Pyke, who at that time was a member of the British Military Mission in Washington, DC. Pyke told me when I visited him that presently the German U-boats were sinking our cargo ships more rapidly than they could be replaced. The principal reason for this was that there were not enough landing facilities between Newfoundland, Iceland, and the northern tip of Europe for aircraft that could locate and eventually destroy the enemy submarines. The construction of new aircraft carriers was slow because of a steel shortage, and those that were built had to go to the Pacific, where our Navy needed them desperately against the Japanese.

Pyke wanted to use flat icebergs as "unsinkable landing fields" for our planes. Small planes had occasionally and accidentally made emergency landings on such places, and the take-off and landing of military aircraft from frozen lakes and river surfaces was a routine operation in Canada. The problem was the fragility of the ice. It was established that a small bomb dropped on a sheet of ice would not just produce a hole, but shatter it entirely.

Pyke wanted to know if it were possible to reduce the brittleness of ice substantially. I told him that when I worked in a pulp and paper mill in Canada, we found that the addition of a few percent of wood pulp greatly increased the strength of a layer of ice. It was agreed that, at our institute, we should find out by how much a few percent of wood pulp or even sawdust would decrease the brittleness of ice.

Walter Hohenstein, a member of our group, started immediately in the deep-freeze house of a Manhattan meat packing company to prepare sheets of ice 3 × 3 feet and 1–3 inches thick, using different amounts of sawdust and different ways to distribute it in the frozen sheet. The mechanical measurements (tensile modulus and impact strength) were carried out by several technical assistants under Alfrey's supervision. Rather interesting results were obtained, and a larger prototype of a "floating landing strip" was in construction when, after the victorious sea battle at Midway in June 1942 and the catching up of our steel production, the project was concluded in September. Its results, however, have been put to good use ever since in all permanent constructions (roads, air strips, bridges, and habitats) in Arctic and Antarctic regions.

Weizmann Institute

Toward the end of the war, in the summer of 1945 when the outcome was already decided, my main concern was to make the work and our institute internationally known. I planned to initiate and sponsor the founding of similar institutions that would constitute a network of polymer research centers cooperating as closely with each other as the prominent schools of organic chemistry did in Oxford and Munich.

Starting in 1942, Meyer Weisgal, an ingenious organizer, fund raiser, and manager, interested a number of wealthy businessmen in England and America in celebrating the 70th birthday of Chaim Weizmann in November 1944 by creating the Chaim Weizmann Institute of Scientific Research in Palestine. This did not require much persuasion, and the necessary funds were quickly raised.

During the years of World War II, Dr. Weizmann spent much time in the United States accompanied by his personal assistant Josef Chon and by Ernst Bergmann, his indefatigable long-time scientific associate. I had met Ernst Bergmann in 1921, when we were both assistants at the First Chemical Institute of the University of Berlin, working under Wilhelm Schlenk. Originally an organic chemist, Ernst Bergmann was to develop into an all-round scientist of unusual competence, a man whose boundless energy and unfailing intuition were responsible for his prominence in the Palestine scientific and industrial communities. Soon after Pearl Harbor, Bergmann began to work at the Polytechnic Institute of Brooklyn on certain steps in the production of synthetic rubber, at that time a high priority in the United States. This afforded me the privilege of seeing him almost daily and of holding frequent conversations with Dr. Weizmann, who visited our laboratory often to discuss Bergmann's work and the entire Rubber Reserve Project. Another topic to receive much attention was the expansion of the Daniel Sieff Research Institute of Science to the Weizmann Institute that was to grow around it.

Planning the Organization of the Institute

Our many conferences and informal chats led to the concept that the institute's first stage should be a nucleus, comprising the most important disciplines and so designed that future expansion would be natural and easy. The models I personally preferred were the Kaiser Wilhelm Institutes in Berlin-Dahlem, founded 30 years earlier, which consisted of institutes for physics, chemistry, biochemistry, and botany. This nucleus later developed into some 30 research centers that now belong to the world-famous Max Planck Gesellschaft. Both Weizmann and

A visit to the Weizmann Institute, 1949. Left to right: me, M. Weisgal, K. Makauer, and E. Bergmann. Dr. Chaim Weizmann is seated on the lower right.

Bergmann knew these institutes very well and had had many contacts with their illustrious directors, Richard Willstätter, Fritz Haber, Carl Neuberg, R. O. Herzog, and Herbert Freundlich. Gradually the following pattern emerged.

The first departments to be organized would be mathematics with an emphasis on applied branches, including aerodynamics and the new discipline of cybernetics; physics with an emphasis on solid-state physics and on the rapidly developing branch of nuclear physics; chemistry with an emphasis on organic chemistry, polymer chemistry, and technology; and biochemistry that stressed molecular biology and genetics.

For each of these departments an advisory body would have to be selected. When possible, a prospective head would already have been approached who, together with the advisory body, would be responsible for estimating the personnel, space, and equipment required for the department. A planning committee was organized to initiate and coordinate these numerous activities. Weizmann asked me to serve as head of this planning committee. He said: "You worked for 6 years at various institutes in Berlin-Dahlem and for 6 years as Research Manager of the largest German chemical corporation; for 5 years you were director of the First Chemical Institute of the University of Vienna, and later you were responsible for the research and development of the Canadian International Paper Company. All of this should give you enough experience and sufficient patience to bring our planning to a successful end." I accepted the appointment on the understanding that I would be privileged throughout to enjoy close cooperation with Ernst Bergmann. In fact, our cordial association in matters having to do with the Weizmann Institute (and with Israeli science and technology in general after the State of Israel was proclaimed in 1948) started then and continued until Bergmann's death in 1975. Much work lay before us and, inevitably, each projected department posed different problems.

C. L. Pekeris of Princeton University declared himself ready to head the mathematics department. He had significant contacts with all of the leading schools of mathematics in the United States, particularly with J. R. Oppenheimer, and he used these to present an excellent and realistic proposal for the size and structure of his future department.

Conditions were more complicated with physics. Here we enjoyed the expert and benevolent advice of James Franck and Isidor I. Rabi. Clearly, future developments in nuclear and high-energy physics would be particularly important and interesting. However, the "Manhattan Engineering Project", where most such work went on at that time, was shrouded in secrecy because of the efforts to develop an atomic bomb. This part of physics at the Weizmann Institute would have to wait, but another important branch—solid-state physics—could be included. Here we had I. Fankuchen, the internationally renowned X-ray crystallographer who also, albeit tentatively, agreed to come to Palestine and take over the solid-state physics division. Later he changed his mind and stayed in Brooklyn. Nonetheless, he rendered invaluable service during the planning and procurement negotiations.

The chemistry department would obviously be Bergmann's own domain. He persuaded Louis Fieser of Harvard and his wife, Mary, to join our advisory group. They had for years maintained excellent relations with most of the world's leading organic chemists. For advice on the polymer division of this chemistry department we turned to George Goldfinger, and W. P. Hohenstein declared his readiness to help us in the buildup of polymer technology. David Rittenberg of Columbia University and K. G. Stern of Brooklyn Polytechnic also consented to join forces in creating a plan for the related departments of biochemistry and biology.

Within 2 months each group had drafted a blueprint for its domain, including the main direction of initial scientific activity, the necessary space, and equipment, and had even established a list of the most desirable co-workers. Several joint meetings followed, some in the presence of Weizmann, in which we tried to coordinate the demands of the individual disciplinary groups while closing gaps and eliminating overlaps.

Planning for the Building and Equipment

Finally a provisional plan emerged for a four-story building that would contain some 40 laboratories and 10 offices, with a staff of some 60 scientific workers plus appropriate technical help.

Weizmann modified and improved various aspects of our report and decided that the design for the new building and its surroundings could only be established in Rehovot with the aid of local architects and authorities. Consequently, he suggested that the members of the planning committee travel to Rehovot and arrive at a final decision regarding the building on the spot, after consulting scientists of the Daniel Sieff Research Institute and the Hebrew University of Jerusalem.

In Brooklyn we were anxious to use the time before the trip to initiate procurement of equipment for Rehovot. Meyer Weisgal's campaign had by now provided the funds necessary for this purpose. Standard laboratory equipment was ordered through Brooklyn Polytechnic from normal suppliers, packaged, and stored for shipment in our stockroom. Highly specialized apparatus, however, had to be handled individually with the supplying companies. The most important items of this type were an ultracentrifuge, an electrophoresis apparatus, an electron microscope, an infrared and Raman spectrometer, and a complete X-ray diffractometer and chemical analyzer.

For the lengthy and sometimes difficult negotiations, I appointed a special procurement group that included I. Fankuchen, W. P. Hohenstein, K. G. Stern, and Ben Siegel for the electron microscope, not only because these instruments were so expensive (more than $250,000) but also because of the very long delivery times (up to 16 months).

All this was well under way in the summer of 1946 when the entire planning committee journeyed to Rehovot, most by boat, some by plane. There it continued its work, having established contact with such distinguished scientists as M. Frankel and A. Farkas and their associates, Aharon Katchalski, Ephraim Katchalski, and Michael Szwarc. This resulted in stimulating some of the younger scientists to join the staff of the new institute.

As the contours and dimensions of the Weizmann Institute became more precisely defined, so did the problem of financing its construction and operation. We were all awed by the magnitude of the funds required. Fortunately, Meyer Weisgal—who was to become the institute's moving spirit, later its president, and finally its chancellor—was at our side. His remarkable enthusiasm and ingenuity kept pace with the increasing demands.

Other International Associations

University of Calcutta

During the war, another promising and important opportunity to sponsor polymer science occurred on an international level. In 1944 Santi Palit, a member of an Indian delegation in Washington, DC, came to work at our institute to familiarize himself with the new science of polymers. He worked here for a year. Eventually, after the war Palit returned to Calcutta, where he started, at the University of Calcutta, a laboratory for the synthesis and characterization of organic polymers. During the next two decades he visited our institute a few times and invited several professors of our team to visit his institute at the University of Calcutta. I was there three times during the 1940s and 1950s, gave lectures, and met a few prominent Indian scientists such as C. V. Raman and N. Batnaghar. For our polymer group, it was a very important early relationship with scientists of a great Asian country.

University of Liège

The real breakthrough in the establishment of international relations between our newly created institute and the rest of the chemical institutions of the world occurred in 1946 when I was invited as a visiting professor to Liège in Belgium. A. Guillet had been their dean of sciences for several years. He had worked on the structure and characteristics of coal, which, of course, is a kind of polymeric material, though not quite as clearly defined as oil. Coal is an important raw material to which Guillet had applied modern physical methods such as chromatography, infrared absorption spectroscopy, X-rays, and others. Because our materials were partly synthetic substances such as nylon and other thermoplastic and thermosetting resins, Guillet felt that I could give his staff a number of lectures that might be of value in reorganizing and enlarging a department of polymer science.

W. H. Dore and G. Smets, also in Liège at that time, were organic chemists who had been quite interested in the investigation and the study of natural polymers, particularly polysaccharides and polypeptides. Thus in the fall of 1946 I went to Belgium and started to give a series of lectures on the general topic of polymer synthesis and on the synthesis and investigation of polymeric materials. At the end of the course Guillet collected all of our lectures in a relatively large volume on the present state of our knowledge of polymeric materials. Some of the other lectures were given by T. Alfrey, A. V. Tobolsky, and R. B. Mesrobian. I had purposely taken with me these younger collaborators and colleagues because I thought that it would be a good opportunity for them to get in touch with the atmosphere of another institute where polymers were beginning to be studied intensively.

International Union of Pure and Applied Chemistry

Beyond the local interest in polymer science in Liège and in Belgium, I was trying to establish contact with official international organizations. The International Union of Pure and Applied Chemistry (IUPAC) had had divisions for about 15 years in such fields as organic chemistry, inorganic chemistry, physical chemistry, analytical chemistry, and biochemistry. Sooner or later, I intended that a division on polymer chemistry should be organized and should keep in contact with laboratories all over the world.

The best opportunity to start such an enterprise is, of course, a conference or meeting that brings together a number of prominent representatives of the field. Such a meeting took place in 1947 in Liège. Important practitioners of organic and polymer chemistry attended, and the meeting was, in fact, the real start of international polymer science. H. S. Taylor came from Princeton, Harry Melville from Cambridge, E. K. Rideal from London, and W. T. Astbury from Leeds. France was represented by George Champetier and Adolph Chapiro from Paris and F. Sadron and F. Latort from other French universities. From Holland came H. Staverman, R. Houwink, P. H. Hermans, and J. J. Hermans; and G. Smets and V. Desreux came from Belgium. There were also other professors from Liège and a few others from the western countries. The war with Germany had ended only 2 years earlier, and it was still difficult to invite German scientists to an international meeting. Two years later this was no longer the case, and then the international circle of scientific contributors to an IUPAC polymer division became even larger.

Food and Agricultural Organization

Contacts with the Food and Agricultural Organization (FAO) were established toward the end of the war by Eugon Glesinger. As director of the forest products branch of FAO, he was responsible for taking stock of existing worldwide forest reserves and working out a long-range plan for a future forestation strategy and for the best possible use of the resources.

Canadian Forest Products Institute

During my stay in Hawkesbury, I had been a member of the Canadian Forest Products Institute. I had worked on cellulose since 1922 and had even written a book, *Physics and Chemistry of Cellulose*, in 1932. As a result, I had had considerable contact with cellulose chemists and engineers all over the world, such as H. Hibbert and L. Purves in Canada; E. Jahn and E. A. Heuser in the United States; W. Hirst and W. N. Haworth in the United Kingdom; E. Berl and S. Bergius in Germany; E. Hagglund and H. Ertman in Sweden; and Professor Namurti in India.

For the first two conferences, in 1949 and 1951, I organized two textile meetings, one in Appleton, Wisconsin, and the other in Dera-dun, India. The subjects to be discussed at these conferences were, essentially:

- Whole-tree technology, that is, utilization of all parts of a tree such as stem, branches, leaves, bark, and roots for the production of energy.
- Pulping of tropical wood species, with emphasis on Brazil, Africa, Indonesia, and India.
- Separation of cellulose from lignin with the aid of organic chemicals, particularly alcohol and butanol.

All the papers presented at these conferences were published in the *Transactions of the Food and Agricultural Organization* of the United Nations. They have contributed greatly to international cooperation in wood science and technology all over the world.

U.N. Industrial Development Organization

There was close contact and intense cooperation with another agency of the U.N., namely the United Nations Industrial Development Organization (UNIDO), which was responsible for technology transfer from the industrialized Western nations to the developing countries. This cooperation started soon after the war and was essentially under the supervision of R. Rothblum, M. C. Verghese, F. May, and K. Youssef.

The respective scientific and technical advisors from our institute and our group were E. Braunsteiner for fibers, films, thermoplastics, and other resin products; W. P. Hohenstein for textile technology and its developments, particularly cotton; G. Oster for dyestuffs, printing, and copying; and A. V. Tobolsky for elastomers and reinforcements.

Dr. Braunsteiner, who had spent several years in Burma, India, Taiwan, Africa, and Brazil, very successfully handled about a half dozen important scientific projects in the domain of her competence. She developed several elaborate reports that contain in full the results of her activities.

The assignments of the other staff members were shorter but equally successful. Many valuable and long-lasting contacts resulted for our institute.

The Institute of Polymer Research and Recent Years

1947–1989

One day in 1946, Dean Kirk took me to the office of our president, Harry Rogers, and said: "Why don't we simply give a name to something that has grown upon us? Let us call it our Institute of Polymer Research." As usual, he had the correct perspective on recent events. Polymer chemistry was becoming a new branch of organic and physical chemistry. It needed and deserved the attention that would be cultivated through an organization devoted to teaching and research.

This situation was familiar to me; it was more or less a duplication of what I had attempted to accomplish at the University of Vienna more than 10 years earlier, on a much smaller scale and with much less opportunity for expansion. Surprisingly, the clear recognition in the New World of the appearance of this new discipline was not an obvious consequence of what had happened during the last two decades. At least not in the academic world! Industry, on the other hand, was fast in sensing novelty and utility. All larger chemical companies established research laboratories, a development that had beneficial consequences for our graduates. Each of them moved from thesis work into attractive and promising positions.

It took more than 10 years before other universities—in Cleveland, Amherst, Akron, and North Carolina—started to initiate the teaching of polymer chemistry and to endorse polymer research on a fully organized scale. However, many distinguished scientists in several locations cultivated with great success, certain special approaches to polymer science and provided strong stimulation and guidance for our own work.

Developing the Teaching Program

In our teaching program, we first followed the schedule developed in Vienna. Gradually, as new branches emerged for additional specialties, we expanded and rounded out the curriculum. I taught two courses, Physical Aspects of Organic Chemistry and Introduction to Polymer Chemistry.

A formal portrait, 1950.

An important part of our activities was the organization of Saturday morning symposia on topics related to polymers. Polymer science was then a new field, generating much enthusiasm. A few prominent scientists from abroad had either moved to the United States before the war (such as Albert Einstein, Peter Debye, James Franck, Richard Courant) or visited soon after the end of the war (such as Max von Laue, W. Bragg, E. K. Rideal, G. Champetier, G. Smets, and C. V. Raman). We established Saturday morning seminars in which these celebrities acted either as chairmen or as featured speakers. The talks were always followed by the most lively discussions. The topics and persons were always very attractive, and visitors came from universities, industry, and government laboratories as far south as Washington, DC, as far west as Chicago, and as far north as Boston. After a few years these Saturday Symposia became famous, generating a high level of unsolicited publicity and providing an important feather in our scientific and educational cap.

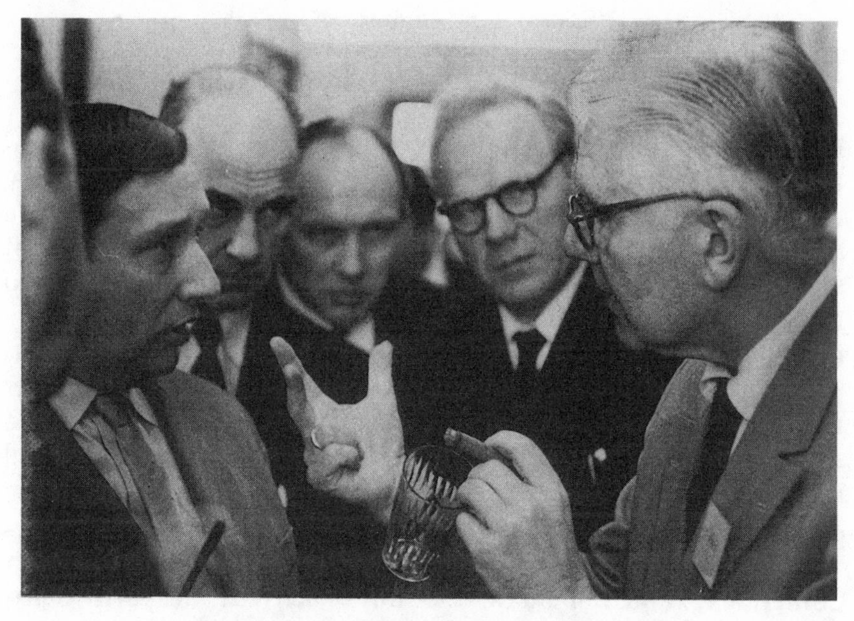

With Soviet scientists visiting Brooklyn, 1955.

> [A] vehicle for closer cooperation between the department and industry was his establishing "Special Summer Courses". The courses, which were the result of a scheme by Mark and I. Fankuchen, served an ancillary purpose. The funds obtained by collecting fees from the attendees were placed in a special travel fund reserved for the faculty. A typical course consisted of a limited number of visiting scientists who remained at the Polytechnic for a period of one or two weeks. While there, they were given intensive theoretical and experimental instruction in a special area of discipline. The first courses were taught in the summer of 1943 by Mark and a number of his associates, including Fankuchen and H. S. Kaufman (X-ray crystallography), and T. Alfrey, P. M. Doty, and B. H. Zimm (molecular weight).[1]

The exposure of our students to leaders in polymer research did much to increase the stimulating atmosphere of our program. Once a year I gave a seminar called "What Is New in Polymers", in which I recounted what I had learned during my extensive travels, in both university and industrial laboratories.

Determining the Direction of Research

As for research, I was anxious to continue our Vienna investigations on copolymerization. This led to the formulation by Alfrey and Goldfinger of the relation between the composition of a monomer mixture and the copolymer derived from it. Later it seemed that the molecular architecture of polymers might lead to materials with interesting useful properties. I encouraged Alfrey to explore, with his students, ways to prepare so-called graft polymers, in which a chain backbone consisting of monomer residues of one type carries side chains of a different polymer. Another research program in our laboratory that led to a useful industrial process was the development of suspension polymerization by W. P. Hohenstein.

I still lecture at ACS meetings and at Gordon Research Conferences, and I spend several days each year at North Caro-

Lecturing at the USSR Academy, Moscow, June 1960.

lina State University as a visiting professor. I am fortunate to be able to continue these activities in my 94th year.

Polymers and the Gordon Research Conferences

The Gordon Research Conferences played an important role in spreading the "polymer gospel". In the beginning, interest was concentrated on fiber-forming polymers. I was fortunate to have the help of Milton Harris of the National Bureau of Standards and of Emil Ott of Hercules. Today, the Gordon Research Conferences have a considerable number of sessions in the polymer field, including topics such as Polymers, Polymer Physics, and Elastomers.

One problem close to my heart was the lack of exposure of chemistry undergraduates to polymer chemistry. It is difficult to understand why, although half of our professional chemists work with polymers, elementary textbooks on organic and physical chemistry hardly mention their existence. I have tried to suggest ways in which this omission might be remedied, but I am afraid we have a long way to go before polymers will receive reasonable treatment.

Patents and Litigation

Shortly after I came to Brooklyn I met Edwin Land. This meeting occurred before his development of instant photography, and his interest was concentrated on the production of Polaroid film. He asked me to advise him on the characterization of the orientation of poly(vinyl alcohol) films and on the relationship between orientation and optical properties that was needed for the drafting of the patent. At the same time I obtained a consultantship with Du Pont on tire cord, which resulted from my contacts with Du Pont while I worked in Canada. The scope of this consultantship broadened greatly as time went on. A little later I was retained by Standard Oil of Indiana to consult on polystyrene, a subject with which I have had a close involvement since my days at I. G. Farben.

The Polaroid consultantship led to my role as an expert witness in a patent suit. We won. Apparently the lawyers were pleased with my performance, and the judge seemed to like my use of a chain of paper clips to demonstrate how polymer molecules behave. As a result, I have been asked to act in a similar capacity in a number of lawsuits in recent years. In the next big suit, which dealt with oil-extended rubber, I represented Goodrich and General Tire; the opposing party was Goodyear and Firestone. The litigation went on for 13 years. I was also involved in the fight over the polypropylene patent. This suit dragged on for 15 years and led to a victory for Phillips Petroleum in the United States, although in all other countries the Montecatini patents were sustained. More recently I testified successfully for O. Wichterle as the inventor of soft contact lenses. Altogether, I have appeared in court as an expert witness some 60 times over the past 40 years.

Becoming a *Geheimrat*

One day, during my early years in Brooklyn, Fankuchen surprised me by addressing me as *Geheimrat*. When I asked him what he meant, he explained that my informal manner was the exact opposite of the "stuffed shirt" behavior of the traditional German professors who had this title of "secret counsellor". This joke of Fankuchen's stuck, and people still call me *Geheimrat* without any idea of what the term stands for. In fact, my style of directing the Institute of Polymer Research was a far cry from the *Geheimrat* tradition. The door to my office was open most of the time, and I encouraged our professors to follow their interest without interference, only pointing out the emergence of promising areas of research.

> *There were so many practical jokes and so much cheerful mischief that the Master (Mark) on his 80th birthday asked himself, "How is it that in spite of it all, something worthwhile came out of it."*—J. Hengstenberg[1]

In 1980, I was one of the recipients of the National Medal of Science. The award was made in recognition of a lifetime of contributions to the development of polymer science. I was very gratified by this award, not so much for myself, but for the recognition that polymer science was now in the mainstream of chemistry. The medal was presented at a ceremony in the East Room of the White House by president Jimmy Carter. The

In 1980, I received the National Medal of Science from President Jimmy Carter for my research on large molecules.

president's science advisor, Dr. Frank Press, served as master of ceremonies. It was a great occasion. My son, Hans, who was President Carter's Secretary of the Air Force, persuaded the president to autograph the picture taken at the ceremony. As you can see, the title that Fankuchen bestowed on me so many years earlier was finally recognized by the highest official in the U.S. government!

I may also have been unusual in that I made no effort to hold on to members of our polymer institute. On the contrary, I went out of my way to facilitate their move to more exciting new opportunities. In this, I may have been influenced by my own experience—Schlenk's advice that I leave his laboratory for the Kaiser Wilhelm Society and Haber's recommendation that I join I. G. Farben. In addition, I felt that it is good for us to have friends in many places, and I was confident that we could replace those who left with new gifted scientists.

Lecturing Around the Globe

> A student once asked Mark, "How do you manage at your age to keep up the pace? You get up early, stay up late, and fly between Europe and America as often as we would take a street car." Mark replied, "You know, this is true. I guess I have no time to get old".[1]

In 1961 Hermann Staudinger visited the United States. I invited him to lecture at the Polytechnic Institute and introduced him as the originator of the polymer concept. This seemed to surprise him, because he could not forget our old controversies concerning the flexibility of polymer molecules. I heard that on his return to Germany he told H. Ringsdorf, his last student: "Now Mark believes in macromolecules!"

Japan. The following year, during my attendance at a meeting in Japan, Professor W. Mizushima (who had been a teacher of the crown prince) arranged that I be invited to give a lecture demonstration to the emperor. This was an exciting occasion for me and perhaps even more so for my wife. I. Sakurada acted as my assistant. Although I knew that the emperor was fluent in

With H. Staudinger at Rockefeller University, 1950.

Enjoying dinner with Professor I. Sakurada of the Institute of Radiation Chemistry in Kyoto, 1966.

English, he had my lecture translated, and all his comments went through an interpreter. Afterward, I was told by the master of ceremonies that it would have been unsuitable for the emperor to speak English because he might have made some mistakes.

During the 1950s Mimi and I embarked upon a tour of South America, starting in Venezuela and continuing into Brazil, Uruguay, Argentina, and Peru, where I lectured and met with prominent scientists from each of these countries.

China. In 1972 I was invited to lecture in China. This invitation happened to bear the same date as the Shanghai Declaration of President Nixon on his historic visit to China. My visit was suggested by Professor Sakurada, the dean of Japanese polymer science, whom I had known for many years. Sakurada was acquainted with Liu Ta Kang, head of the Chemical Institute of the Chinese Academy of Sciences. I spent 3 weeks in China, lecturing and visiting the production facilities of their emerging polymer industry. Professor Goldberg of the Princeton Physics Department was there at the same time. We were the first American scientists to visit the country after the installation of the Communist government.

Europe. In more recent years I have divided my time between Europe (where Vienna is my base) and the United States (where I am mostly in Brooklyn). I first returned to Europe in 1947. Since then I have been abroad about 500 times and have visited all the continents, including Greenland and Antarctica. I still cross the ocean both ways about a dozen times a year. Both personal and professional motives make me spend a good part of my time in Europe. My brother, my sister, a cousin, and their families live in Vienna and I have relatives in Hungary and Rumania. I also have some 25 relatives in Israel. After the war, they were all in a poor condition and needed help, both financial and moral. I still like to visit the surviving family members, altogether some 30, whenever I visit Europe and Israel. In the 1950s and 1960s, around 40 of my World War I comrades still lived in Austria. They would organize dinners, banquets, and dances when I was in the country, and I enjoyed these sentimental get-togethers. Today there are only a few survivors.

The scientific colleagues with whom I had been associated in my youth have spread all over Europe and beyond. Now I visit them to relate American polymer science to the rest of the world, both as an educational activity and as a way to make our institute visible. Fortunately, I can lecture in English, French, Italian, and German, and have some knowledge of Spanish and Russian; this has been very helpful. In this crusade, which is still going on at a reduced intensity, I have visited over 100 countries and over 1000 scientists and engineers.

> Years have not changed Mark. On a European tour in the late 1950s, he visited a particularly attractive campus in France. The hosts took special pleasure in showing him the Institute's spectacular architecture. Tongue-in-cheek, he responded by telling them of the beauty of his own campus at Brooklyn Polytechnic. Brooklyn Poly is, of course, located in the serviceable, but not "ivy walled" downtown of Brooklyn, not far from New York's East River.
>
> This episode does not end here. A few months later, the Frenchmen toured the U.S., and naturally came to Brooklyn to see Professor Mark's campus. In the tradition of his *Geheimrat,* Murray Goodman, then a young Professor at Poly, greeted the visitors, and saved honor by convincing them that the New York Supreme Court Building, a few blocks away, was the building Dr. Mark had described.[1]

These scientific contacts took about half of my time during the early postwar period. Later, several of my colleagues (C. G. Overberger, M. Goodman, Herbert Morawetz, and E. M. Pearce) took on a considerable amount of this responsibility, and

On the occasion of a symposium held in October 1972, in memory of Aharon Katchalsky. Left to right: Murray Goodman, Herbert Morawetz, Gerald Edelman, me, Charles Overberger, and Melvin Calvin.

I cut back travel to a third of my time. I was glad of it because there were a few more or less pressing things to do at home.

Publications

Some of the "home jobs" I had to take care of involved publications. We had some difficulty in having certain polymer papers published in the *Journal of the American Chemical Society.* I visited A. A. Noyes to get his advice on the creation of a polymer journal. Noyes was encouraging, although he did not want the ACS to be involved. Then, in 1945, I persuaded M. Dekker and E. Proskauer to have Interscience Publishers launch *Polymer Bulletin*, containing mostly work carried out at Poly. The reaction of the polymer community was favorable, so we started publication of the *Journal of Polymer Science* in 1946. At that time contributions were only sufficient for six rather small issues per year! I also felt that a monograph on the mechanical behavior of high polymers was badly needed. Alfrey undertook the task of writing a book on this subject, the first of its kind.

The *Journal of Polymer Science* needed an improvement in its contributions and a shortening of publication delays from 10–12 months to 6–8 months. The "High Polymers" monograph series, which I edited, was a high priority; so was the "Abstracts Series" on Fibers, Rubbers, and Plastics, which was doing well in terms of subscribers but not so well in speed of appearance. We then started to think about an American version of the German *Ulmann.* With the aid of Dean Kirk and Donald Othmer, I persuaded Interscience Publishers to publish an *Encyclopedia of Chemical Technology,* which was edited by Kirk and Othmer. Three editions of this encyclopedia have now been published. I am editor of the *Encyclopedia of Polymer Science and Engineering,* which is now in its second edition. Altogether, the five encyclopedia editions amounted to about 100,000 printed pages and netted John Wiley, their publisher, several millions of dollars.

Family

For the first 7 years of our marriage, 1922–1929, Mimi and I abstained from having children. My jobs were not secure

Celebrating my 50th Birthday party in 1945, Brooklyn. Mimi is standing behind and to my right; I am seated before the cake. Sons Hans (taller) and Peter, in matching jackets, are in the back row, far right. Also present are Dean R. Kirk (seated far left), W.P. Hohenstein (standing with pipe), and I. Fankuchen (seated far right).

enough, and we did not want to take any chances. Our health was good and we could afford to be patient. Then, in June 1929, a boy was born, with some difficulties during birth because he was a large child. Mimi nursed him for a few months and he (Hans Michael) grew up in our elegant apartment in Mannheim, Germany—Augusta Anlage 22. Two years later another boy arrived (Peter Herman), whose birth was very easy. The two boys grew up together peacefully and without serious childhood diseases. Mimi and I were thinking of a third child (hopefully a girl), but when my work in Ludwigshafen came under political pressure, we decided to wait.

In Vienna Hans entered public school in 1935 at 6 years of age. Conditions were not much better, and we gradually decided that the two boys were enough for the difficult years of the near future.

Relaxing at our home on Ocean Avenue in Brooklyn, 1964.

The excitement of our emigration and settlement in the small mill town of Hawkesbury were highly enjoyed and appreciated by both boys. Eventually they attended the Ecole Populaire in Hawksbury, where the language was French. Mimi was fully occupied with fixing the house and running the kitchen; we had three servants, a gardener–chauffer, a cook, and a maid. The Canada days were short. In 1940 we moved to Brooklyn, into a relatively large apartment at 325 Ocean Avenue. After his high school graduation in 1947, Hans went to the University of California in Berkeley. Two years later, Peter entered Harvard University.

In 1943 we bought a summer home at a lake north of New York, near the small city of Peekskill. For almost 30 years this home offered inexpensive and very pleasant weeks of relaxation.

By the mid-1960s Mimi, unfortunately, became affected by angina pectoris. At first the symptoms were mild and she could lead essentially a normal life. But later her condition deteriorated. In October 1968 I took her back to Vienna, where we first stayed in the Hotel Europe and later (January 1970) in

With son Hans and Edward Teller at my 70th birthday party, 1965. Teller was Mark's student at the Technical University in Karlsruhe in 1926 and 1927.

Three generations; with grandson Rufus and son Hans.

the St. Joseph and the Hartmann hospitals. After several weeks of terrible suffering, my beloved Mimi died on March 10, 1970. Four years later my brother, Hans, one year younger than I, died of a stroke. The last severe and bitter loss was the passing away of my younger son, Peter, in 1979 at age 48 after a very painful and lengthy case of lung cancer.

Mimi and I use Egypt's local transportation on a visit to the Sphinx, October 1966.

My Philosophical Outlook

My philosophical approach to life gained an early and strong foundation in high school (1906–1909), where the classical and humanitarian spirit of Greek and Latin thinking was the backbone of our education. Later, in the philosophical part of my education, I was strongly influenced by Karl Prinz, who lectured on Kantian philosophy. The essence of all of this was that you cannot get anything in life (position, influence, or economic security) without working for it consistently in such a manner that all your motives could be useful principles of a successful social order. Truth, dependability, benevolence, and the absence

Demonstrating the structure of chain molecules, Brooklyn, 1967.

of malice would have to be the pillars of my existence. Very often it was difficult to find them at all. Even when, in any given case, they were attainable, it was not easy to follow them through. Thus, life was a delicate compromise between the things I should have done and those that I actually did. Scientific and professional life in the laboratories, in the classrooms, and at seminars and meetings was relatively easy to handle. However, the treatment of people (scientists, executives, and government representatives) was sometimes so complicated that I never knew whether I was doing the right thing or not. There was only one simple overriding rule: keep cool and never lose your composure!

References

1. Stahl, G. A. In *Polymer Science Overview: A Tribute to Herman F. Mark*; Stahl, G. Allan, Ed.; ACS Symposium Series 175; American Chemical Society: Washington, DC, 1981; Chapters 1, 3, 6, and 10.
2. Schlenk, W.; Mark, H. *Chem. Ber.* **1922,** *55B,* 2285, 2299.
3. Gonell, H.; Mark, H. Z. *Physik Chem.* **1923,** *107,* 181.
4. Mark, H.; Weissenberg, K. Z. *Physik* **1923,** *17,* 301.
5. Mark, H.; Weissenberg, K. Z. *Physik* **1923,** *16,* 1.
6. Mark, H.; Weissenberg, K. Z. *Physik* **1923,** *17,* 347.
7. Hassel, O.; Mark, H. Z. *Physik* **1924,** *25,* 317.
8. Hoffmann, H.; Mark, H.; Z. *Physik Chem.* **1924,** *111,* 321.
9. Mark, H.; Pohland, E. Z. *Krist.* **1925,** *61,* 293.
10. Mark, H.; Pohland, E. Z. *Krist.* **1925,** *61,* 532.
11. Staudinger, H. *Chem. Ber.* **1920,** *55,* 1073.
12. Polanyi, M. *Naturwissenschaften* **1921,** *9,* 288, 337.
13. Polanyi, M. In *Fifty Years of X-ray Diffraction*; Ewald, P. P., Ed.; Oosthoek's: Utrecht, Netherlands, 1962.
14. Sponsler, O. L.; Dore, W. H. *Colloid. Symp.* **1926,** *4,* 174.

15. Polanyi, M. Z. *Physik* **1922,** *9,* 123.

16. Staudinger, H.; Heuer, W. *Chem. Ber.* **1930,** *63,* 222.

17. Katz, J. R. *Naturwissenschaften* **1925,** *13,* 411.

18. Pauling, L.; In *Polymer Science Overview: A Tribute to Herman F. Mark;* Stahl, G. Allan, Ed.; ACS Symposium Series 175; American Chemical Society: Washington, DC, 1981; pp 95–99.

19. Mark, H. *Scientia* **1932,** *51,* 405.

20. Mark, H.; Susich, G. V. Z. *Physik* **1930,** *65,* 253.

21. Bergmann, E.; Mark, H. *Chem. Ber.* **1929,** *62B,* 750.

22. Hengstenberg, J.; Mark H. Z. *Krist.* **1929,** *70,* 283.

23. Mark, H.; Wierl, R. *Naturwissenschaften* **1928,** *16,* 725

24. Mark, H.; Wierl, R. Z. *Physik* **1929,** *53,* 526; *55,* 156; *57,* 494.

25. Fischer, E. *Chem. Ber.* **1907,** *40,* 1754.

26. Freudenberg, K.; Braun, E. *Liebigs. Ann. Chem.* **1928,** *460,* 288.

27. Freudenberg, K. *Liebigs. Ann. Chem.* **1928,** *461,* 130.

28. Meyer, K.; Mark, H. *Chem. Ber.* **1928,** *61B,* 593.

29. Meyer, K.; Mark, H. *Chem. Ber.* **1932,** *61B,* 1932.

30. Meyer, K. H.; Mark, H. *Chem. Ber.* **1928,** *61B,* 1939.

31. (a) Mark, H.; Wierl, R. *Naturwissenschaften* **1930,** *18,* 205, 753, 778; (b) Mark, H.; Wierl, R. Z. *Physik* **1930,** *60,* 741.

32. Alfrey, T., Jr.; Goldfinger, G. *J. Chem. Phys.* **1944,** *12,* 205.

33. Mayo, F. R.; Lewis, F. M. *J. Am. Chem. Soc.* **1944,** *66,* 1594.

34. Guth, E.; Mark, H. *Monatsh. Chem.* **1934,** *65,* 93.

35. Kallman, H.; Mark, H. *Naturwissenschaften,* **1925,** *13,* 1012.

36. Kallman, H.; Mark, H. Z. *Physik,* **1926,** *36,* 120.

Additional Reading

Morawetz, H. *Polymers: The Origins of Growth of a Science;* Wiley: New York, 1985.

Mark, H. *The Use of X-rays in Chemistry and Technology;* J. A. Barth: Leipzig, Germany, 1926.

Meyer, K. H.; Mark, H. *Der Aufbau der Hochpolymeren Substanzen;* Hirschwaldsche Buchandlung: Berlin, 1930.

Meyer, K. H.; Mark, H. *Der Aufbau der Hochpolymeren Organischen Naturstoffe;* J. A. Barth: Leipzig, Germany, 1930.

Mark, H. "Polymer Chemistry in Europe and America: How It All Began"; *J. Chem. Ed.* **1981,** *58,* 527.

Staudinger, H. *Die Hochmolekularen Organischen Verbindungen, Kautschuk und Cellulose;* Springer: Berlin, 1932.

Reflections by Two Eminent Chemists: Drs. H. Mark and P. J. Flory; American Chemical Society: Washington, DC, 1982. (Videotape)

Meyer, K. H.; Mark, H. *Hochpolymere Chemie;* Akademische ver-Pagsgesellschaft: Leipzig, Germany, 1937.

Meyer, K. H. *Natural and Synthetic High Polymers;* Interscience: New York, 1942.

Mark, H. *Das Schwere Wasser;* F. Deuticke: Leipzig, Germany, 1934.

Fankuchen, I.; Mark, H. *X-ray diffraction and the Study of Fibrous Proteins;* Interscience: New York, 1946.

Mark, H.; Tobolsky, A. V. *Physical Chemistry of High Polymeric Systems;* Interscience: New York, 1950.

Index

Index

Copy editing: Colleen P. Stamm
Production: Peggy D. Smith
Indexing: Janet S. Dodd

Production Manager: Robin Giroux

Printed and bound by Maple Press, York, PA